CW00521434

Increased pressure for fast and decisive responses to complex problems is a challenge for all decision-makers and analysts. This book from Gilbert Probst and Andrea Bassi provides excellent tools for public policy makers as well as business leaders and academics, on how to deal with complex problems. The step-by-step approach followed by examples and case studies, gives the reader excellent guidance into the author's method.
**Borge Brende, Minister of Foreign Affairs, Norway**

In a world of increasing uncertainty and velocity, *Tackling Complexity* is making sense of complexity and brings clarity, insight and actionable ideas. Through a series of tools including systems mapping, stakeholder analysis, scenarios and decision-making protocols, the authors show how today's complex world can be understood and acted upon. It's a must-have toolkit for everyone who works in a complex world.
**Lynda Gratton, Professor of Management Practice,**
**London Business School**

As global volatility escalates, and organisations and individuals across sectors become more and more interconnected, leaders will be increasingly burdened by difficult decisions in complex systems of stakeholders. *Tackling Complexity* deftly outlines a practical and systematic approach for solving these challenging problems.
**Dominic Barton, Global Managing Director, McKinsey**

Understanding and dealing with complexity is essential when it comes to modelling business situations and strategizing. Applying the 5 steps in my strategy processes throughout my career proved most helpful. This approach allows us to truly recognise the interdependences, include different stakeholders, understand the impact and timeframe of interactions and make stronger long-term decisions. The compelling cases, the structured guide, and systemic perspective lead policy makers and strategists to better decisions.
**Dr Peter Fankhauser, COO, Thomas Cook**

Simple, linear decision making belongs to the previous century. Economies and firms are now integrated across the bounds of jurisdiction function and impact. The pity is that too few decision makers are equipped to deal with complexity. This phenomenal contribution by Professor Gilbert Probst and Dr Andrea Bassi places the techniques of complex decision making within reach of every executive, regardless of whether he/she is in the public or private sector.
**Trevor Manuel, Minister of the National Planning Commission, South Africa**

What bedevils decision makers today is not just the speed of change, the increasing interconnectedness of decisions or the risk of unintended consequences. It is all of the above! In this smart, savvy book, Probst and Bassi offer a practical roadmap for today's executives to tackle truly complex, systemic challenges. It is a breath of fresh air and an antidote to the overly simplistic formulae that account for so much of the management literature.
**Professor Rita Gunther McGrath, Columbia Business School**

Our world has shrunk. A simple metaphor explains how. When humanity lives in 193 separate countries, it is no longer like living in 193 separate boats. Instead, we are all living in 193 separate cabins on the same boat, but with no captain to take care of it. Hence, this book could not be more timely. On this new global boat of ours, we need to apply systems of thinking to save our planet. Global challenges require global solutions. This book explains how to produce them.
**Kishore Mahbubani, Dean of the Lee Kuan Yew School of Public Policy, NUS and author of The Great Convergence**

Systemic changes ... require new forms of leadership from men and women – and especially women – willing to be the vanguards of change. For them, *Tackling Complexity* provides an invaluable route map of what it takes to drive change and succeed in the VUCA [Volatile, Uncertain, Complex, Ambiguous] world that is undoubtedly here to stay.
**Paul Polman, CEO, Unilever; Vice-Chairman, WBCSD (From the Foreword)**

*Tackling Complexity* serves as a valuable resource in using systems thinking for strategy and policy development using the multi-stakeholder approach. The book focuses on the main challenges at the heart of global and local development, highlighting key consequences, and proposes methods and tools to better understand and address complexity in order to improve effectiveness and increase sustainability.

The greater the complexity of the system, the greater the risk of systemic breakdown – yet also the greater the opportunities for transformation. *Tackling Complexity* demonstrates the foresight and spirit of solidarity needed to strengthen the opportunities related to organisational learning, and the transformation towards a more resilient system.

**Klaus Schwab, Founder and Executive Chairman,**
**World Economic Forum (From the Foreword)**

We are now recognising that the challenges to the public's health are not solved by single, simplistic solutions – just as we previously learned that the world is not flat. This outstanding book offers a toolkit that brings the insights of systems thinking and analysis to the creation of effective solutions – whether for global epidemics of obesity or diabetes, or anticipating the health impacts of climate change.

**Linda P. Fried, MD, MPH, Dean, Mailman School of Public Health,**
**Columbia University**

# Tackling Complexity
A Systemic Approach for Decision Makers

# TACKLING COMPLEXITY

## A SYSTEMIC APPROACH FOR DECISION MAKERS

GILBERT PROBST AND ANDREA M. BASSI

Published by Greenleaf Publishing Limited
Aizlewood's Mill
Nursery Street
Sheffield S3 8GG
UK
www.greenleaf-publishing.com

Cover by LaliAbril.com

Printed in the UK on environmentally friendly, acid-free paper
from managed forests by CPI Group (UK) Ltd, Croydon

British Library Cataloguing in Publication Data:
A catalogue record for this book is available from the British Library.

ISBN-13: 978-1-78353-081-6 [hardback]
ISBN-13: 978-1-78353-082-3 [PDF ebook]
ISBN-13: 978-1-78353-083-0 [ePub ebook]

# Contents

# Foreword

**Paul Polman**

The acronym VUCA—Volatile, Uncertain, Complex, Ambiguous—may have its origins in the military, but it is increasingly clear that it applies to all aspects of our lives today. The fact is we operate in an age of fast-moving and increasingly unpredictable change. No one country, society, industrial sector or individual organisation is immune. We are all impacted.

Navigating this new reality is made even more challenging by the increasingly interdependent nature of today's world. The issues and predicaments we face are linked inextricably as never before. There is no better or more dramatic illustration of this than the food, water, energy and climate nexus, so effectively highlighted over recent years by the World Economic Forum and others. How do we guarantee food security for a rapidly rising population in the face of growing water and energy constraints, many of them directly attributable to climate change? No wonder one leading scientist has warned of a 'perfect storm' of global events.

Increasingly, business has found itself in the eye of this storm, mistrusted by large sections of society and seen, with some justification, as part of the problem and not part of the solution to many of today's challenges. This has to change. Business can no longer afford to be a bystander, content to sit on the sidelines doing the minimum necessary to acquire its 'licence to operate'. The challenges require—and citizens demand—a different approach. Permissible growth in the future has to be based on sustainable and equitable models. Having acquired a licence to operate, it's time for business to earn a licence to lead.

The military and defence officials who first identified the VUCA world used to speak of the need for 'burden sharing', for sovereign nations to spread more evenly the responsibility for defending freedoms around the world. It's a military parallel that also has resonance today. For its part, business has to accept a much greater share of the responsibility for everything that goes on within the length and breadth of its value chain. Putting your own house in order is a necessary—but insufficient—condition for developing sustainable growth models.

This is the essence behind the Unilever Sustainable Living Plan, which gives expression to the company's commitment to double its size while reducing its overall environmental footprint and increasing its positive social impact. It's a plan that covers the entire value chain, from sourcing to manufacturing to the way consumers use our products. This requires the kind of holistic, systems-based thinking that Andrea Bassi and Gilbert Probst argue needs to become mainstream.

If we get it right we can make a difference. However, the sheer size and scale of the challenges we face means that even the largest and most internationally dispersed businesses and organisations are limited in the scope of what they can achieve. System-wide changes rely on a critical mass of interested parties, all willing to enter into deep partnerships and collaborations, founded on new levels of trust and a commitment to action not debate.

There is still a long way to go. We are far from reaching the tipping point that is needed, though the tide is certainly turning, in my view. The commitment to put an end to illegal deforestation and develop sustainable alternatives for commodities like palm oil and soy, for example, is an inspiring illustration of what can be achieved when governments and industry partners come together determined to bring about transformational market-wide changes.

None of this is easy. Everything carries a risk. Taking the first step is often the most difficult. It takes courage and a willingness to focus on long-term horizons, not short-term results. Systemic changes therefore require new forms of leadership from men and women—and especially women—willing to be the vanguards of change. For them, *Tackling Complexity* provides an invaluable route map of what it takes to drive change and succeed in the VUCA world that is undoubtedly here to stay.

*Paul Polman is CEO of Unilever, and Vice-Chairman (and next Chairman) of the World Business Council for Sustainable Development*

# Foreword

**Klaus Schwab**

We are living in the most interdependent era in history; the world today is changing very fast, and everywhere. In contrast to the past, leaders today are faced with a tremendous range of global issues—economic, political, environmental, societal and technological—which intersect with each other through a highly complex web of causal links. In such conditions, it is no longer possible to assert, for example, that an economic problem will be confined to the economic sphere, or that an environmental risk won't have repercussions reverberating across different domains.

These dynamics have created an environment that only a decade ago was the province of futurists, but today the ubiquity of these forces are having a profound impact on decision makers' understanding of the world and their ability to exercise leadership. Operating in an increasingly complex environment requires a greater degree of contextual intelligence and social agility. In practical terms, this means that no leader can afford to think in 'silo' terms, but rather must be much more holistic and adaptive in order to 'connect the dots'.

Leaders can only adapt to increasing interdependence if and when they are highly connected and inclusive, engaging with all those who have a stake in the issue at hand. Ultimately, this is the principle embedded in the World Economic Forum's 'multi-stakeholder theory', as neither governments, business nor civil society can resolve these issues alone. It is only through the collaborative efforts of all stakeholders of global society that true progress can be achieved. It is this entrepreneurial collaboration in the global public interest that embodies the true spirit of the World Economic Forum—an institution that understands the nature of complex systems all too well, as it mirrors the world at large.

In order to efficiently address problems, the foremost experts in the fields of academia, business, government, international organisations and society must be better equipped to address complexity. *Tackling Complexity* serves as a valuable resource in using systems thinking for strategy and policy development using the multi-stakeholder approach. The book focuses on the main challenges at the heart of global and local development, highlighting key consequences, and proposes methods and tools to better understand and address complexity in order to improve effectiveness and increase sustainability.

The greater the complexity of the system, the greater the risk of systemic breakdown—yet also the greater the opportunities for transformation. *Tackling Complexity* demonstrates the foresight and spirit of solidarity needed to strengthen the opportunities related to organisational learning, and the transformation towards a more resilient system.

*Klaus Schwab is the Founder and Executive Chairman of the World Economic Forum*

# Preface

This book provides a step-by-step approach to using **systems thinking** in order to solve complex problems. It proposes a technique with which to better understand the problems, the context in which they arise and the tools required to directly inform each step of the decision making process.

Practically, this approach can be used to identify problems, analyse their boundaries, design interventions, forecast and measure their expected impacts, implement the interventions, and monitor and evaluate their success/failure.

The main innovation introduced is that systemic thinking emphasises the system, which is made up of interacting parts, rather than prioritising events that need immediate fixing.

The book touches on global issues related to strategic management, as well as issues related to sustainable development in the public and private sectors.

In these contexts, systems and their complexity can be investigated by means of indicators, influence tables, causal diagrams, scenarios and simulation. The use of a multi-stakeholder approach leverages the strength of these tools, making them far more effective.

The targeted audience includes business leaders and their strategic foresight professionals, government officials involved in strategy/policy-making and their support staff, professionals working with international organisations (e.g. the UN system), students and academics.

Systems thinking implies the consideration and analysis of many different views and stakeholders. In this sense, we have greatly profited from conversations with several colleagues and peers. They have supported the preparation of this book in several ways. Our thanks go to Selima Benchenaa, Reuben Coulter, Fernando Gomez, Niccolo' Lombardi, Stephan Mergenthaler, Vanessa Candeias and Maribel Adame Valero. Our thanks go also to the various private and public sector leaders that have helped us shape the contents of the book through fruitful collaborations during the last few years.

We would also like to acknowledge the financial contribution of the Foundation for the Advancement of Systems-Oriented Management, St Gallen, Switzerland.

*Andrea M. Bassi and Gilbert Probst*
*Geneva, 2014*

# 1
# Introduction

## 1.1 Purpose of and rationale for dealing with complexity

Our socioeconomic systems continue to grow and evolve. We need to acknowledge that, consequently, our decisions often fail—they are ineffective and create unexpected side-effects. The speed of execution is increasing constantly and markets and systems respond almost immediately, making decision making challenging. There is little or no room for failure.

> Since 2008, the high volatility of the financial markets has burdened national leaders with considerable pressure. They have to make crucial decisions on monetary and fiscal policies and have to constantly monitor the markets. Simultaneously, they have to improve private sector access to credit (in the context of declining demand) and higher lending (although unexpected losses have heavily impacted financial institutions) to promote economic growth. They also have to have a wise monetary policy to stimulate exports, as well as monitor inflation and speculation.

> During the last decades supply chains have become increasingly globalised. Product manufacturing is now often internationally and strategically located to minimise production costs. Consequently, since the supply chain is mostly far from its customers, any shock

in terms of raw material availability (primarily sourced locally), labour availability, productivity, energy prices, etc. would have direct consequences on production costs and consumer demand. To counter these risks, companies such as PUMA have started to determine their global footprint to improve their resilience and use the unique strengths of their network of suppliers.

The emergence of multiple, interconnected crises exacerbates the recent challenges. We have recently experienced financial, energy and food crises, among others, and the world is feeling the effects of climate change. The growing global interconnectedness of our society, economy and environment is no longer questioned. This interconnectedness invariably leads our thoughts to the 'butterfly effect'.[1]

There is increasing evidence that the economic growth of the last few decades has been achieved at the expense of natural capital. Further, there is growing recognition that the over-exploitation of the same resources that fuelled economic growth in the past causes the current crises (UNEP, 2011a). In fact, while value has been created by the use and transformation of natural resources, stocks have been greatly depleted. For example, only 25% of commercial fish stocks—mostly low-priced species—are now underexploited (FAO, 2008) and 27% of the world's marine fisheries had already collapsed by 2003 (Worm *et al.*, 2006). Oil production has reached its peak and is declining in most countries (EIA, 2009). With the current water supply predicted to satisfy only 60% of world demand in 20 years, water stress will increase (McKinsey & Company & 2030 Water Resources Group, 2009). Chemical fertilisers have boosted agriculture yields (FAO, 2009a) to the detriment of soil quality (Muller & Davis, 2009). Between 1990 and 2005, 13 million hectares of forests disappeared per year (FAO, 2009b), the size of Bangladesh and Greece.

No wonder the general public and policy-makers still believe that the goals of economic growth and environmental protection, as well

1 In chaos theory, the butterfly effect is the *sensitive dependence on initial conditions*, where a small change at one place in a deterministic nonlinear system can result in large differences to a later state.

> as national and energy security, involve a complex set of trade-offs (Howarth & Monahan, 1996; CNA Corporation, 2007; Brown & Huntington, 2008). But tools that support a systemic analysis can shed light on the dynamic complexity of the social, economic and environmental characteristics of our world. The goal should be to evaluate whether new system-wide strategies can create synergies and help the world move toward resilient economic growth, job creation, low-carbon development and resource efficiency.

The rapidly evolving environment in which we live requires faster and more decisive responses—responses to what appear to be multiple and simultaneous challenges—leaving little room for a careful analysis of alternative intervention options. Such pressure on decision makers can lead to rushed decisions. These decisions are taken on the basis of recent events. They do not consider the complex dynamics underlying the true causes of the problem. Short-term reporting or rewarding practices often motivate the search for immediate solutions and lead to unintended consequences or side-effects. These may exacerbate the problem in the medium and longer term.

Our systems have always been complex, but global crises make complexity evident to everyone at a level that cannot go unnoticed. They highlight planetary and human boundaries, which, once pushed or surpassed, can lead to unexpected consequences on other—apparently disconnected—parts of the system.

This book clearly distinguishes between complicated and complex systems. The analysis focuses predominantly on the latter.

- **Complicated systems** are composed of many different interacting parts whose behaviour follows a precise logic and repeats itself in a patterned way. They are therefore predictable. Automatic watches with mechanical movements composed of hundreds of coordinated elements are examples of complicated systems. Another example is the manufacturing process of a car during which several actions (steps) occur in a well-planned sequence by means of several work stations.

- **Complex systems** are dominated by dynamics that are often beyond our control. These dynamics are the result of multiple interactions between variables that do not follow a regular pattern. However, their dynamic interplay can lead to unexpected consequences. Society is a complex system driven by emotions (the human component), infrastructure and our environment. For instance, not all drivers take the same road to work every day. The choice of road depends on personal needs (do we have time?), the information we receive and pay attention to (has there been an accident on the road?), the transportation mode utilised (car, bus or metro) and our interest in the environment (e.g. in $CO_2$ emissions). All these factors never work in exactly the same way every day, but there is always a specific rationale behind every decision. The system is therefore complex, not chaotic or complicated.

Currently, globalisation has made complexity equally relevant for governments and the private sector.

- **Complexity is relevant for public policy-makers** because most countries face several concurrent challenges that simultaneously affect (and are affected by) social, economic and environmental dimensions. To reach any stated goal and to shift closer to a sustainable development pathway, we need decisive policy interventions that will simultaneously support socioeconomic development and environmental conservation.

Qatar is an example of a country that paid renewed attention to complexity when designing its national development policies. In the 2009 Qatar Human Development Report, the government recognised that the country was at a crossroad. Its development models, which were based on fossil fuel exploitation, needed to also consider future challenges related to oil price fluctuations, environmental degradation, the expected population growth, and equitable access to services and opportunities. According to the report, 'sound management of Qatar's hydrocarbon resources will continue to secure improvements in standards of living. However, an improved [economic] standard of living cannot be the only goal of a society' (*Government of Qatar & UNDP, 2009*).

- **Complexity is relevant for the private sector** because the business environment is changing rapidly. The patterns of demand and supply from emerging countries are evolving, technology is developing rapidly, and energy and natural resource prices are highly volatile.

The manufacturing sector is confronted with complex challenges. The most important of these is to maintain a high level of competitiveness in increasingly interconnected markets. Even the definition of competitiveness changes rapidly. On the whole, it depends on the supply side—the speed of production, the costs and the innovation—as well as on the demand side—the perception of the quality of the product or service—and its brand or image. Manufacturers therefore need to comply with global standards and expectations. They have to keep pace with their competitors while simultaneously adapting their products to local needs and cultures.

This book offers decision makers in the public and private sectors the steps and tools they need to address current complex (or wicked) issues. Informed decision makers will then have the capability to analyse challenges with a systemic perspective, which will lead to lasting solutions that sustainably improve the performance of their system.

Decision makers will immediately identify with Churchman's definition of 'wicked problems': 'that class of social system problems which are ill-formulated, where the information is confusing, where there are many clients and decision makers with conflicting values, and where the ramifications in the whole system are thoroughly confusing' (Churchman, 1967). This complexity may frighten decision makers, or they may be unable to appreciate and understand it. They therefore tend to oversimplify reality and merely address the apparent and most obvious causes of a problem. This is a linear approach, according to which there is only one possible solution to a given problem. However, linear thinking is based on a series of logical mistakes that lead to a number of common errors when attempting to deal with complex (or wicked) problems.

Throughout the book, a step-by-step approach guides the reader to identify these problems, analyse their boundaries, understand their interconnections, design interventions, forecast and measure their expected impacts, and monitor and evaluate their success or failure (see Table 1).

| Decision-making phases | Conceptual mistakes | What to do | Steps |
|---|---|---|---|
| 1. Problem identification | #1 Abundance of data allows us to find ultimate solutions and predict system behaviour | Delimit the problem, identify the causes and the effects | Define the boundaries of the problem<br>Identify the causes and effects<br>Analyse future behavioural paths and impacts |
| 2. System characterisation | #2a Every problem is a direct consequence of a single cause<br>#2b We only need an accurate 'snapshot' of the actual state of the system to find solutions | Map the complexity and explore the dynamic properties of the system | Build a causal diagram and review the boundaries of the system<br>Create a shared understanding of the functioning of the system<br>Identify key feedback loops and entry points for intervention (strategy/policy identification) |
| 3. Strategy/policy assessment | #3 The problem will be solved with the implementation of the intervention selected | Identify the 'learning' capabilities of the system | Design potential interventions<br>Assess interventions (anticipate gaps and early warning signals)<br>Select viable intervention options and indicators |
| 4. Decision making and implementation | #4 With a problem-oriented optimisation, the solution will maximise benefits for all | Evaluate the proposed solution using different perspectives and assess the impacts across sectors and actors | Use a multi-stakeholder approach to assess roles and responsibilities<br>Analyse the expected impacts across sectors and actors<br>Define the strategy and action plan |
| 5. Monitoring and evaluation | #5 Monitoring and evaluation do not affect the decision-making cycle, they only evaluate the system performance | Assess the effectiveness of the implemented interventions and the system responses to redefine the top priorities and the need for further action | Implement the strategy and monitor the development of the system<br>Analyse the impacts across sectors and actors<br>Use lessons learned for the next decision-making process |

**TABLE 1 Conceptual mistakes, solutions and steps for an effective decision-making process using systems thinking**

Each chapter starts with an analysis of the drawbacks of conventional approaches to problem solving, examining the weaknesses and limitations of decision-making models based on linear thinking, and relying on accurate predictions of the future (Probst, 1985; Probst & Gomez, 1989; Probst & Gomez, 1995).

It then proposes a decision-making approach that contextualises issues by considering their social, economic and environmental drivers and impacts. The advantages of a systemic approach are described from a strategic point of view that emphasises the opportunities related to organisational learning and the transition to a better performing and resilient system.

The book provides relevant examples to highlight the flexibility of the approach and the proposed tools. These include indicators, influence tables, causal loop diagrams and scenarios. The tools have been chosen for their ease of use in a multi-stakeholder context, as well as for their effectiveness in analysing complex problems.

## 1.2 Audience and user guidance

The targeted audience includes business leaders and their strategic foresight professionals, as well as government officials involved in strategy/policy-making and their advisory staff, professionals working with international organisations, as well as students and academics interested in systems analysis.

Throughout the book, each step of the decision-making process proposes and analyses issues related to strategic management, as well as to the challenges related to policy-making for sustainable development (e.g. in the context of the green economy, climate mitigation and adaptation; see for instance UNEP 2009), in order to illustrate the concepts and the steps better.

The book provides generic guidance, or a method that can be applied to develop effective interventions. The end goal is to improve resilience and increase sustainability. The reader may therefore want to utilise the stages and techniques described in this book in his or her respective context to improve the performance of his or her company, sector or country.

The book's structure follows the decision-making process (see Figure 1). It places possible solutions within a cycle that includes: 1) the definition of issues (or agenda setting) in a systemic context; 2) system characterisation; 3) decision making; 4) implementation; and 5) evaluation. This is done to ensure that: 1) policy or strategic issues are appropriately identified and defined; 2) potential solutions are formulated and 3) assessed; 4) the solution that increases synergies and reduces trade-offs is chosen and implemented; and 5) the adopted solution is monitored and evaluated.

Chapter 3 introduces the process's first stage, the problem identification, which identifies trends in the present and potential future challenges. Chapter 4 focuses on the system characterisation phase, which identifies possible intervention options to solve the problem. Chapter 5 addresses the strategy and policy assessment phase, which evaluates the feasibility of different intervention options, as well as their effectiveness within and across sectors. In Chapter 6, the decision-making and implementation process is analysed, focusing on the key steps regarding refining and operationalising the strategy and policy. Finally, Chapter 7 describes the monitoring and evaluation phase, which assesses the actual system response and evaluates the efficiency and effectiveness of the interventions to inform the next round of the decision-making process.

Each chapter starts with the presentation of common conceptual mistakes in dealing with complex problems: systemic mistakes we human beings make all the time, traps we fall into because we do not understand or respect systems. The chapter then proposes key steps to identify and implement solutions (see Figure 1 and Table 1). The main tools that support the analysis of complex problems are presented in the Annex.

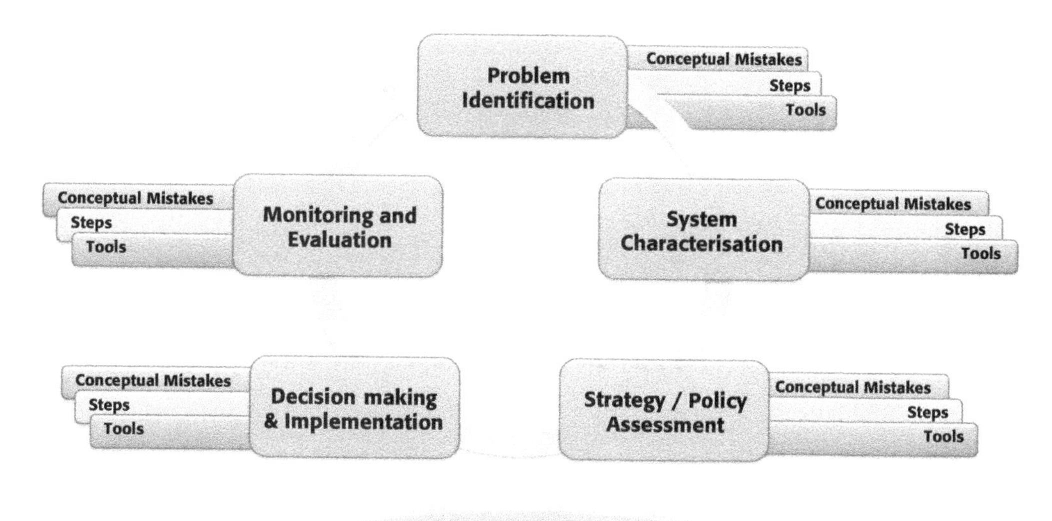


FIGURE 1 The decision-making process, also adopted as the basic structure of this book. Each chapter includes the presentation of conceptual mistakes, steps, actions and tools for problem solving

# 2
# Case study
## Why a systemic approach matters

Among the many concurrent crises of the last few years, the food crisis is particularly relevant and recurring. Despite the industrialisation of the agriculture sector's value chain and the countless advances made in the field in the last century, its relevance and fragility remain undisputed (IAASTD, 2009; FAO, 2009c).

Society is critically dependent on agriculture. It provides food in several forms and shapes, contributing directly to life. In addition, working in the sector is the only subsistence option for billions of people. Consequently, about 40% of the world's jobs can still be found in the primary sector—the vast majority of which are in developing countries (Hurst, 2006; World Bank, 2008). However, the agricultural sector's contribution to the economy has greatly and constantly declined over time; currently, it is only 3% of the global GDP and 25% of low income countries' GDP (World Bank, 2013).

Recent megatrends have greatly affected the agriculture sector—positively and negatively. First, driven by innovation and investment, agricultural productivity has increased substantially. On the other hand, soil quality has gradually declined over time. The improvements in productivity have therefore often been obtained through innovation that has been offset by a decline in the quality of the environment (Bassi *et al.*, 2011). But the limits of improvement are being reached, also due to additional stresses, such as those related to climatic changes. These changes

**Set the direction**
- Establish and enforce consistent, transparent regulation to attract investors
- Increase funding for agricultural development, especially infrastructure and research
- Open trade policies that facilitate market access for developing countries
- Ensure rural access to education, healthcare, and capital—regardless of gender
- Lead stakeholders in holistic transformations

**Innovate and invest**
- Develop and scale interventions that are proven to meet the combined objectives of the New Vision
- Increase access to agricultural finance through innovative risk - sharing partnerships
- Step up engagement in holistic transformations

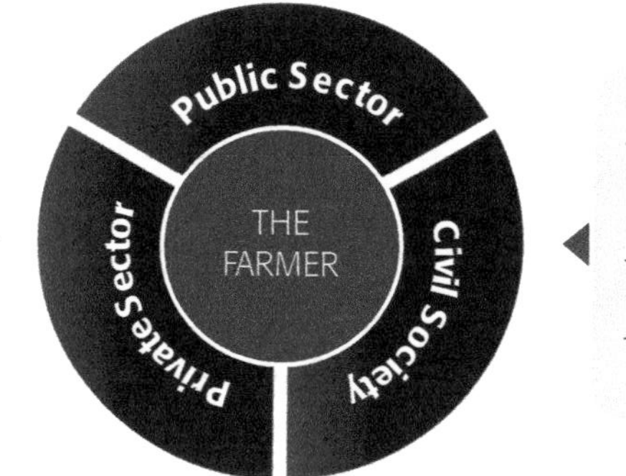


**Mobilise the community**
- Actively represent the voice of citizens, communities, and the environment in holistic transformations
- Train and organise local producer organisations
- Leverage capital to bridge gaps in the value chain and reduce risk

FIGURE 2 The role of the public and private sectors and civil society in support of the New Vision for Agriculture

*Source*: World Economic Forum, 2010

are: increasing soil erosion, precipitation variability, and exacerbating pests and diseases. These challenges and the globalised world's fierce open market competition have led to a decline in the sector's profitability in various parts of the world. The regions that are most vulnerable to climate change and countries with no or little infrastructure to support effective market access are especially affected. Further, the underdevelopment of rural areas has led to millions of people migrating to urban areas in search of better jobs and income. This megatrend has triggered the emergence of additional environmental stresses, thus further exacerbating the challenges that the agriculture sector—especially in developing countries—faces.

In light of the above challenges and of the projected growth in food demand, there have been several calls for action to decisively transform the agriculture sector.

The World Economic Forum, in partnership with several private sector companies and international research institutes, developed a *New Vision for Agriculture* (World Economic Forum, 2010). This vision calls for a multi-stakeholder approach in which the public sector sets the direction with policy interventions, civil society mobilises the community, and the private sector innovates and invests. This integrated approach can support farmers effectively with the end goal of *providing food security for all in an environmentally sustainable way while generating economic growth and opportunity.*

A specific experience of the Bühler technology group provides an excellent case study that illustrates a strategic process required to address the present and upcoming challenges in the agriculture sector from a private sector point of view.

**The Bühler management was aware that more than 2 billion people worldwide suffer from malnutrition, or an invisible hunger for micronutrients.**

The Bühler group designs machines to process dietary staples such as wheat, maize, cacao and rice, and partners with builders that install their machines. The CEO of Bühler had a specific goal in mind. He wanted to turn one **problem** that characterises nutrition into an opportunity for his company and for consumers. Specifically, the Bühler management was aware that more than 2 billion people worldwide suffer from malnutrition, or an invisible hunger for micronutrients such as vitamin A, folic acid, iodine, iron and zinc. The lack of these nutrients can have very serious and irreversible effects. For example, one person goes blind every four minutes due to a lack of vitamin A. Further, iron deficiency can stunt children's mental and physical development without it being clinically apparent, while an iodine deficiency can trigger skeletal malformations, speech disorders and hardness of hearing. More broadly, in developing countries, but also in certain industrialised countries, this 'hidden hunger' reduces adults' vitality and capability, thus negatively affecting their productivity (Kennedy *et al.*, 2003).

Bühler discovered that the cause of this type of malnutrition is frequently an unbalanced diet of starchy, high-energy foods such as rice, wheat or maize. It then investigated the contribution it could make to solve the problem. In this respect, the **goal** of the management was to design and develop technology to support the reintegration of nutrients, thus increasing the nutritional value of the food, reducing waste and adding economic value to the 'enhanced' product.

With this goal in mind, the Bühler management developed a **strategy** based on technology development that would lead to a win–win–win situation for the company, its clients and consumers. The strategy was based on three main pillars: 1) the product choice; 2) research and technology development; and 3) market analysis and a dissemination plan.

**The *goal* of the management was to design and develop technology to support the reintegration of nutrients, thus increasing the nutritional value of food, reducing waste and adding economic value to the 'enhanced' *product*.**

1. **Product choice: rice.** White rice is the staple food of more than 2 billion people—an immense market potential for nutritional upgrades. Further, white rice is produced when a grinding machine removes the outer layer of a whole grain rice—the rice bran. However, the grinding process leads to the loss of most of the vitamins and minerals. Polished, white rice does provide consumers with the necessary calories, but not with sufficient micronutrients. This means that Bühler, with its knowledge and

**The *strategy* is based on three main pillars: (1) the product choice, (2) research and technology development, and (3) market analysis and a dissemination plan.**

capabilities that matched the production process of white rice well, was very well placed to play an important role in the market.

2. **Technology development: a new white rice production process.** Bühler's Nutrition Solutions business unit worked with the chemical group DSM to develop a solution. After four years, the technology that reconstructs new rice grains by extracting additional vitamins and minerals from rice flour—a by-product of the milling process—was ready. The business plan was that Bühler would deliver the machines and DSM would provide rice millers with NutriRice™ micronutrients.

3. **Dissemination plan: develop a product identical to white rice and use markets and channels in China with which Bühler is already familiar.** Internal studies indicated that, despite its better nutritious value, Asian consumers would not tolerate too much broken rice. The product was therefore designed so that its form, colour and consistency cannot be distinguished from natural grains, which would make it acceptable to Asian consumers. Further, Bühler was already working with Chinese rice millers. Banking on the expertise it had accumulated over time and the good working relationship, the company decided to approach them with the new technology.

Considerable effort went into devising this strategy because the project, which would lead to a new business model, was very important for Bühler. The company had always demonstrated high technological competence, now it was time to offer specialised knowledge of food ingredients and their usage. This project was expected to contribute to the expansion of Bühler's expertise.

Further, many signs pointed to the success of this project. First, the economic argument: when sold as animal feed, broken rice is sold for a much lower price than its true economic and nutritional value. With the use of the new process, the product not only has more value and is enriched with vitamins, but the broken rice is also upgraded. The rice miller can therefore expect to increase its revenues and profitability by upgrading broken rice grains. Since the reconstituted and fortified rice grains are mixed with natural rice at a ratio of 1 or 2 per 100, consumers would absorb sufficient nutrients that would otherwise be lost through the production process and when cooking.

The implementation of the strategy had very surprising results. The project failed. In hindsight, the goal clearly required a systemic approach, but the strategy was linear. More specifically, Chinese millers did not want to produce and commercialise enhanced rice grains and concerns were raised about food manipulation. The project reached an impasse even before it could approach the final consumers. Even worse, Bühler and DSM had already made considerable investments in developing and testing the technology, and in elaborating various business plans.

**The project was a failure. In hindsight, the goal clearly required a systemic approach, but the strategy was linear.**

With the same goals in mind and still convinced of the project's potential, the Bühler management carefully analysed all the problems that had emerged during the implementation of the first strategy. They started elaborating a new, more customised and systemic approach. Specifically, the **new strategy** included a partnership for the 'in-house' production of micronutrients. A market was created through marketing and product differentiation. In addition, extensive communication with legal authorities led to the full certification of, and a patent for, the product and the production process. By taking a systemic and multi-stakeholder approach, the new strategy considers the local context better and creates the necessary enabling conditions for the product to be a success.

**Bühler and DSM designed and implemented a *new strategy*. By taking a systemic and multi-stakeholder approach, the strategy considers the local context better and creates the necessary enabling conditions for the product to be a success.**

The implementation of a new strategy should therefore take the following into consideration:

- **Monitor implementation problems and customise the approach**. The biggest challenge that comes from implementing a strategy emerges from the culture in which the change is to be made. Whether it is change in the culture of an organisation, or in the cultural characteristics of a nation, change always targets individuals' attitudes and behaviour. Bühler and DSM found themselves confronted with a cultural 'otherness'. Attempts to avoid uncertainties and control the future would therefore be far more difficult in China than they would be in Switzerland or Germany. Hofstede and Bond, researchers of intercultural differences, summarised the Chinese attitude as follows: 'What is true or what is right is less important than what works' (Hofstede & Bond, 1988). The rice millers thought the new product was simply too strange to allow them to predict its success. They considered the risk too high and did not buy the NutriRice™ technology. Customising a strategy to the local context is thus very important for the ultimate success of a project.

- **Invest capital to ensure the project start and access to consumers.** The most immediate challenge was the rice millers' unwillingness to sell the product—an unexpected system-driven response that would also make complete sense from a Western point of view. Their unwillingness was partly due to the timing of the product launch, which coincided with the melamine scandal in China. Milk products had been illegally enriched with melamine, a poisonous substance used in the manufacture of plastics. Consequently, food additives frightened Chinese consumers greatly. The rice millers therefore did not want to enhance China's most important staple food. Another problem was linked to the traditional labour agreement. Rice millers had always provided the technology that transforms whole grain rice into white rice. They found the process of reconstituting rice grains by enriching them with 'chemical additives' simply too strange. The Bühler and DSM team considered the relevance of the external events even more powerful than the internal ones. They thus started exploring scenarios and alternative strategies to break the rice millers' resistance. With a better understanding of the market, they decided to gently introduce the product to the rice millers. The most successful entry point that they identified was a joint venture with Wuxi NutriRice™ Co. Ltd. A joint venture would allow them to produce NutriRice™ grains instead of relying on rice millers. NutriRice™ could also be directly introduced to the end customer, which would allow them to identify potential early warning signs for large-scale commercialisation.

- **Diversify the 'look and feel', be transparent and create a product segment.** With access to the end consumer, the project team gained various additional insights. First, it became clear that the customers wanted the NutriRice™ grain's colour to differ from that of other rice. This made the 'manipulation' of the rice grains more explicit and transparent, which was highly appreciated under the circumstances. Further, the customers only believed in the NutriRice™ grains' improved nutrition if they could distinguish them from other rice. It is therefore clear that, irrespective of how comprehensive one's market research is, one can never fully anticipate how customers in a new market will behave. No one could have predicted that the Chinese would prefer coloured NutriRice™ grains, when all rice grains are usually separated according to colour, which is the benchmark for the assessment of their quality. Had this been known prior to production—through market surveys and other means of reaching out to customers—the development costs spent on imitating a rice grain perfectly could have been saved.

- **Ensure active communication with all stakeholders.** Excellent communication is key to ensuring that changes are implemented and markets evolve. In most transformation processes, it is primarily the workers who have to be won over by a strategy's sense and purpose. Bühler and DSM wanted to win over the Chinese market, but selling complex products—such as enriched food products—is an additional challenge. The product's scientific and innovative background makes it difficult to explain the full extent of its value and to communicate the message clearly and simply.

  One should first serve the major stakeholders, then work alongside them, before finally guiding them in order to communicate effectively in a transformation process. The co-workers, suppliers, clients, investors, regulators and other stakeholders must be persuaded to accept changes and participate in the process. Each stakeholder should be individually approached and all concerns should be considered to ensure effectiveness. In the Bühler and DSM case, communication was especially demanding. They had to explain the details of the product to each of the participants (i.e. the producers, buyers, consumers and even the government) for the market to accept, produce, distribute and sell it.

- **Obtain a formal endorsement and create synergies with public policy.** Bühler and DSM encountered several challenges at the political level. The melamine scandal led the Chinese government to start drawing up new food regulations. Food producers and consumers are likely to be cautious about new food manipulations until these regulations have been approved and implemented. The Chinese government required an independent body to prove the amount of vitamins and minerals present in NutriRice™. Initially this was a barrier, but it also proved to be a powerful synergy. Showing trust in the local institutions and research facilities, Bühler and DSM allowed them to take on this task. Despite measurement challenges, the analysis resulted in positive nutritional evaluations. Obtaining a patent for the new process also proved to be a great hurdle in China. Granting two foreign companies a patent to modify their most important staple food was a delicate situation for the Chinese authorities. The final challenge was to encourage the government to cooperate in order to disperse NutriRice™ across the country. This process can take years and the magnitude of the challenge should not be underestimated. On the other hand, the end goal of the project should always be kept in mind. In this case, improving nutrition goes well beyond the product's market success, and policy support is essential.

- **Design and implement a dynamic business plan and adapt to the changing market.** The new business model includes the sale of rice grains to rice millers and selling NutriRice™ directly to the market. Two options were subsequently developed for the medium and longer term. The first was that rice farmers had to accept the product and continue buying NutriRice™ grains from the joint venture. The second was that above a certain volume it would become more cost-effective for farmers to produce NutriRice™ themselves. It would then make more economic sense for them to buy the technology from Bühler and produce the grain. However, the rice farmers only had experience with selling a simple staple food product and limited expertise regarding marketing a new product. Bühler and DSM therefore focused on a more sustainable business model for all the participants and provided specific marketing support. This resulted in the joint venture covering its costs after three years of considerable investments. It had also created a dedicated business model for each phase of the market's maturation.

The **result** of the new strategy is a resounding success. The product is natural and innovative, enriched with much-needed nutrients, and has a higher economic value. The new process reduces waste by turning the rice grain into a product with a higher value. Furthermore, the process allows various market actors throughout the value chain to benefit. The strategic process is now systemic, employs a multi-stakeholder approach and is highly customised to the local context and culture. Scenarios have also been used to analyse the potential evolution of the market. In addition, the external drivers are addressed, the impact of decisions is monitored and evaluated, and the business model is continuously adapted to the evolving market conditions (i.e. the strategy and decision-making processes are continuous and circular), all of which reduce the risks and uncertainty. This approach is not only replicable, but has also been applied to wheat and maize. All these elements of the implemented strategy and the result obtained indicate that it is indeed possible to realise a New Vision for Agriculture.

**The *result* of the new strategy is a resounding success. The strategic process is now systemic, multi-stakeholder and customised to the local context and culture. Risks and uncertainty are addressed by means of scenarios, by monitoring and evaluating the impact of decisions and by continuously adapting the business model to the evolving market conditions.**

The Bühler and DSM case study includes examples of typical stumbling blocks that are encountered when a new strategy is implemented. It also shows that using a linear approach can lead to opportunities being missed. On the other hand, it demonstrates that a successful strategic plan can be developed in the context of complexity with rapidly changing global markets.

# 3
# Phase 1
## Problem identification

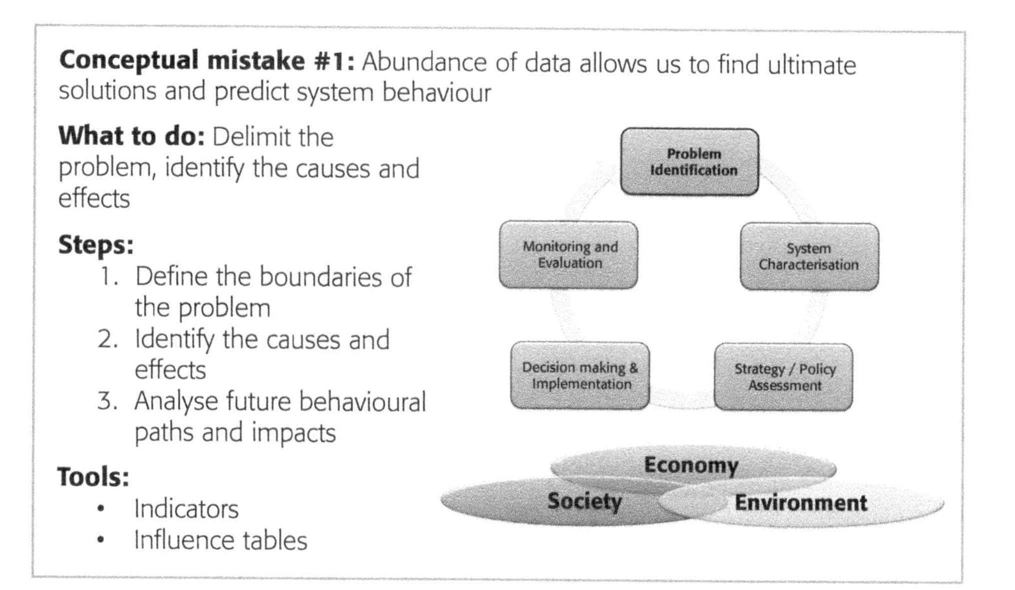

## 3.1 Conceptual mistake #1: Abundance of data allows us to find ultimate solutions and predict system behaviour

Many people think that complexity is inversely related to the availability of quantitative data. This implies that data allows us to debunk complexity and understand it. The assumption is that finding solutions to complex problems only requires detailed information about the presumed direct causes and the harmful effects of the problem (Tversky & Kahneman, 1974; Heinhorn & Hogarth, 1978).

> ***An example:*** the escalating terrorist attacks of the fundamentalist Muslim group Boko Haram in Nigeria are generally reported in the media as part of a terror campaign rooted in the religious differences between the Muslim-dominated north and the Christian-led south. The media provide objective data to show that the terrorist groups have grown in recent years, as have their number of victims and the size of the land that they control. Despite this data, which is essential to frame the problem, the information may not be sufficient and/or adequate to understand the complex dynamics of the problem and how this complexity emerges from the situation (system). For instance, trends and patterns regarding access to political power, land ownership, the formal and informal involvement of foreign governments and multinational oil companies, and tribal value systems should also be considered. If decision makers were to solely take the superficial information—i.e. the available quantitative data—into consideration, they would simply focus on solving religious clashes. Instead, they should explore different possible solutions to address the diverging interests of the various stakeholders.
>
> The smokeless cigarette 'Premier' is an example of a failed business decision due to managers' reliance on objective data. RJ Reynolds introduced 'Premier' in 1988, after the anti-smoking lobby had managed to raise awareness of the negative health consequences of passive smoke. According to then current survey data, the development of a smokeless cigarette could lead to significant profits by building on the new market opportunities that the anti-smoking

campaign offered. However, after 'Premier' had been launched, safety advocates called the product a 'nicotine-delivery system'. Rumours were spread that the new cigarette could be used to smoke crack cocaine. The anti-smoking organisations also accused RJ Reynolds of using the appealing high-tech design of the 'Premier' to encourage young people to smoke. And this wasn't the worst: even smokers did not like the 'Premier'. It had a bad taste and odour, and inhaling was difficult. In general, smokers mistrusted smokeless cigarettes. Ultimately, RJ Reynolds was obliged to withdraw the product from the market after having invested about US$325 million in its development, testing and marketing (Annacchino, 2003).

When collecting and analysing data, attention should be paid to its level of depth and what it potentially explains. We observe events every day. They are the result of all the complex interactions that occur in a system—also considered its tipping points. Events are therefore the result of the patterns and trends that systemic structures generate. Further, the variety and diversity of our mental models—our own understanding of how the system works—mean that various actors in society perceive and interpret systemic structures (e.g. rules, laws and routines) differently. This leads to the creation of micro and macro patterns, which result in specific events.

***An example:*** we can use the 2008 global financial crisis as a broad theme to show examples of the events, patterns or trends, structures, and mental models that played a role in its manifestation. An interlinked series of changes in the system variables–whose complexity made it difficult for analysts to detect the problem prior to its culmination–caused the effects (events) of the crisis. More specifically, the collapse of the Lehman Brothers investment bank was one of the first events to focus the world's attention on the financial crisis. However, in isolation, this event would not provide decision makers with enough information about the real causes of the financial collapse. The bank's collapse was a symptom, not the virus (cause).

The patterns and trends that led to the collapse of Lehman Brothers include the growing concerns about the banking system's stability and the slowdown of the housing market and the economy as a whole—Wall Street started declining in the first quarter of 2008. These trends provide valuable information on the system's direction before the financial crisis emerged.

It is therefore possible to identify worrying trends that led to the financial crisis due to changes in the systemic structures of the banking system. The lack of control of risky financial transactions had led to an overstretched system in which bad loans (subprime) could easily be sold as market products to risk-prone organisations and clients. The growing housing market and increasing housing values, together with the ease with which loans could be obtained, fuelled dramatic changes throughout the banking system. In particular, bankers had overlooked the basic rules of loan schemes—such as the primary importance of guarantees. They thus paved the way for widespread speculation with housing prices. In essence, financial deregulation reached a tipping point—investment banks could operate without external control.

Finally, the different understandings of how the system worked were a key factor in the process that led to the financial crisis. For example, the widespread belief that the housing market and housing values would continue to grow facilitated the increase in subprime lending. Based on this perception of reality, professionals in the sector and homebuyers easily reached agreements regarding loans, both believing that the risk was very low. The increasing housing values would, after all, allow all the parties to re-sell property at a premium and repay the loans. In this case, the mental models were disconnected from reality. The extent of the crisis would not have been so wide if there had not been several layers of actors involved in trading bad loans and making the domino effect possible.

When we fully trust data that merely represents an event, we greatly simplify our approach to decision making and problem solving. This is likely to be ineffective. This approach prioritises the end result, which we believe should immediately and effectively fix the problems that the event caused. Moreover, by following this approach, no effort is made to prevent similar events from happening again in the future. Analysing the process—i.e. the patterns, the systemic structure and the mental models—that resulted in the current situation could prevent a recurrence. However, events are just the tip of the iceberg (easy to see and explore) while the mental models—the underlying structures and patterns—are the core elements of the system that lead to the creation of events (see Figure 3). The evolution and historical interplay of the multiple variables that influence a system are often ignored when we only focus on events (or crises). Nevertheless, they have the potential to provide essential information about the way a system 'learns' and reacts to sudden or gradual changes.

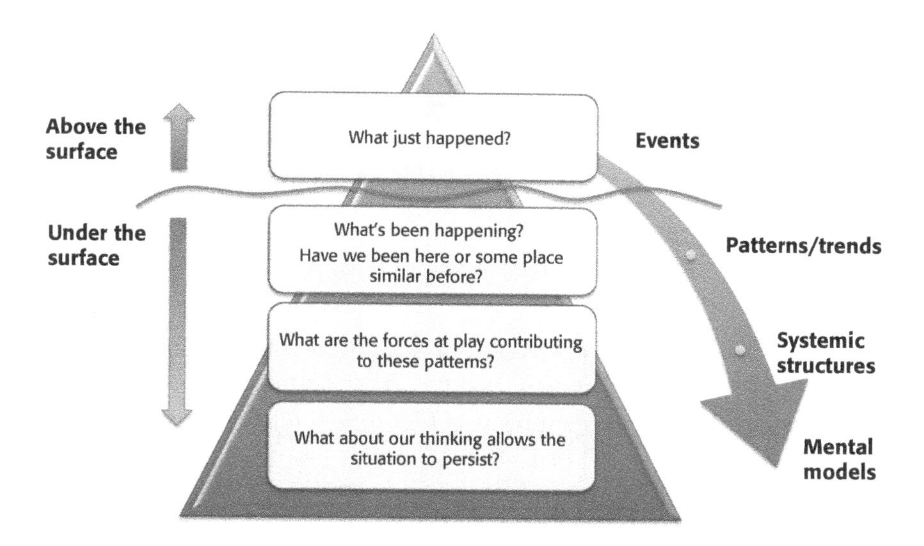


FIGURE 3 The main ways of explaining reality, using the iceberg analogy

***An example:*** the United Nations Environment Programme (UNEP), through its green economy work, provides an example of how misleading or, at least, insufficient objective data can be when a complex problem regarding the link between economic development and the related environmental impacts needs to be solved. Examining the state of the economy and only considering objective and widely accepted indicators—such as GDP—could lead to distorted conclusions. These conclusions could undermine the success of medium to longer-term development strategies. Conversely, for a greener economy and sustainable development, policies are needed that take into consideration that all economic activities occur in the natural, physical world. These activities require resources such as energy, water, materials and land. In addition, economic activity invariably generates material residuals, which enter the environment as waste or polluting emissions (UNEP, 2010a).

Finally, by excessively emphasising data on events, decision makers tend to look at something from one point of view—theirs. They erroneously believe that their perception of reality (their mental model) is universally valid. They ignore the role that the existing interactions play between the systemic structure and the other actors in the system. This approach precludes the creation of a *learning environment* that considers the social, economic and environmental contexts from which the problem emerges.

With this rationale, the same solution may be applied to similar problems in different contexts. This decision is based on the assumption that certain observed conditions will inevitably lead to a determinate situation. In reality, the results are likely to be very different. In other words, linear thinkers accept that 'the chart is the patient' (Gall, 1977).

***An example:*** unsustainable deforestation is a problem that affects many countries and has harmful consequences for the environment, economy and society. Data on deforestation (e.g. deforested area, the protected land) and its direct causes (e.g. wood extraction, the expansion of agriculture land) provides essential, but insufficient, information to allow sustainable solutions to be identified for the problem. The patterns of behaviour resulting in deforestation (e.g. the use of wood for cooking and heating) and the underlying structures responsible for these patterns (e.g. cultural norms and values regarding tree-cutting activities, the perception of forest goods as unlimited resources) could lead to different solutions being adopted. For instance, market-based approaches to counter deforestation (e.g. carbon trading, biodiversity offsets, certification, eco-tourism) have proved successful in some cases, but failed in others, or even exacerbated the problem. This happened with eco-tourism in some areas of India: local governments sold forests to private eco-tourism companies without considering the reaction of the indigenous communities, who are culturally attached to the forests and rely on them for their livelihoods. Consortia of indigenous communities and local NGOs currently strongly oppose public policies that favour eco-tourism development. Further, they accuse tourism companies of being the real cause of ecological damage (Global Forest Coalition, 2008). An in-depth analysis of the system's structure would certainly have revealed these unexpected events.

***An example:*** the establishment of the Euro Disney amusement park in Paris (now known as Disneyland Paris) is another example of unexpected consequences due to an inaccurate analysis of the mental models that influence systemic structures. The company decided to invest heavily in creating the first Disney amusement park in Europe, which was opened in 1992. Based on lessons learned from their other parks in the United States and Japan, the managers decided to increase the ticket prices, ban alcohol—to respect the Disney family values—and prohibit picnics. However, these decisions ignored distinctive French cultural values and habits. First of all, the French happily consume a glass of wine during family outings. Furthermore, bringing their own food to parks is a Europe-wide practice reflecting Europeans' general preference for homemade food, which is also a cost-saving option. The prohibition of picnics and the high ticket prices made Euro Disney too expensive for many families. Consequently, profits after the opening of the park were well below expectations, causing considerable losses (Nutt, 2002). This failure could have been avoided if the decision makers had better analysed the specificities of the 'French system', instead of replicating Disney's business model in other countries and cultures.

## 3.2 What to do: Delimit the problem, identify the causes and the effects

As seen in the previous section, objective event data is not sufficient to analyse the complex—and often hidden—dynamics that underlie the emergence of a problem. It is only through an understanding of the structure of the system that the most appropriate leverage points can be identified. This allows effective action to improve the system performance. In particular, the causes and effects of the problem—as well as the actors affecting and being influenced by it—must be clearly identified and analysed and their relations with the other key drivers of the system have to be considered.

A systemic problem identification exercise generally leads to a list of issues—including potential opportunities, which may be missed without adequate interventions—to which decision makers should pay serious attention at any given time.

The key issues for a national government might include rising food prices, air pollution and illegal immigration. Company issues include erosion of their market share, their reduced competitiveness, employment attrition and their cost control.

Decision makers could pay attention to many issues, but they should seriously concentrate on just one sub-group of prioritised issues. Problem identification is therefore a process in which decision makers—with the help of adequate tools—recognise that certain issues are more important than others. They need to decide which factors and drivers are included in the system and which are not. They need to find out which part of the system generates the problem, as well as which stakeholders, and their relative perspectives, have to be analysed. The goal is to identify and analyse which elements and which stakeholders interact within the system to create the problem. Decision makers base their decision on what to prioritise, which is based on a series of criteria that help them narrow the focus of their attention. In order to 'scan' the system and identify worrying trends, as well as their causes and effects, a logical sequence of actions needs to be followed.

This chapter suggests three steps to guide decision makers through the problem identification phase:

1. Define the boundaries of the problem
2. Identify the causes and effects
3. Analyse future behavioural paths and impacts

### 3.2.1 Step 1: Define the boundaries of the problem

The problem identification phase starts with the definition of the problem boundaries. This action will narrow the focus of the analysis and delimit the area of potential intervention. The first basic principle for successful problem-solving exercises is to exclude all those factors that are not directly related to the problem. It is important to focus on the main causes and effects of the problem and to only analyse the part of the system that contributes to it, whether directly or indirectly. Decision makers should therefore define boundaries that are: 1) sufficiently open to include the essential cause–effect relations that a single causality approach would exclude; and 2) sufficiently narrow to avoid generalisation and a loss of focus.

Qualitative and quantitative data should be used to assess the system performance and to detect anomalies and/or undesired trends. Depending on the system analysed, various types of trends should be considered, not only the declining trajectories. Indicators are valuable tools to support such an analysis. In particular, the interactions between relevant performance indicators are likely to reveal patterns and trends to clarify the relations between the key variables and help define the problem boundaries.

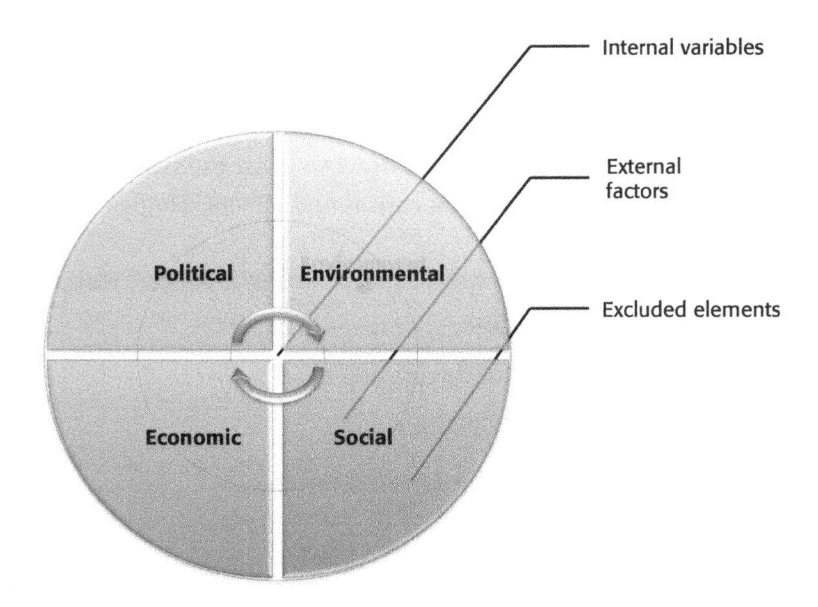


FIGURE 4 **Graphical representation of the boundaries to be considered: social, economic, environmental and political dimensions can be internal (affecting and being impacted by the problem), external (influencing the system and the problem, or its solution) or excluded (not related to the problem)**

***An example:*** the analysis of national statistics may reveal a decline in fossil fuel reserves that the increasing exploitation of renewable energy sources cannot compensate for. Once this worrying trend has been identified, the problem boundaries should be accurately defined. In other words, a distinction should be made between those variables that are in some way related to the problem and those that are not. For example, energy production and consumption patterns, as well as energy intensity and energy-efficiency levels, should be considered within the boundaries of the problem—they contribute to the exploitation of fossil fuel resources. Other elements, such as access to water, are probably irrelevant to this specific issue. They should not be part of the policy formulation process.

***An example:*** the finance sector is starting to recognise the importance of cross-sectoral indicators in designing successful strategies. The UNEP Finance Initiative and its constituents state that the loss of natural capital has direct and widespread negative effects on financial performance. Coordinated strategy and policy intervention are required to counter these systemic risks. The financial markets do not yet understand the risks that companies face from disruptions of vital ecosystems through their supply chains. Only a systemic analysis can highlight an investment and engagement opportunity for investors.

***An example:*** Fairtrade Labelling Organizations International—a network of 19 fairtrade labelling initiatives currently covering 23 countries—is a success story. Fairtrade organisations successfully addressed the problem of the imbalance between farmers' remuneration and the price of the finished products. Volatile commodity prices, inappropriate labour standards, weak farmer trade unions and poor consumer awareness were identified as some of the main causes of the problem. In response, the fairtrade organisations decided to mark their products with a label guaranteeing that farmers—acting through a fairtrade farmer cooperative—have been paid a fairtrade price. This price is based on the average cost of sustainable production rather than on volatile global commodity market prices. The correct delimitation of the problem, including its main causes and effects, allowed fairtrade organisations to find targeted solutions and adapt them to the different contexts in which the problem arose.

**How to identify indicators throughout the decision-making process**

Following the decision-making process, the main steps to identify indicators that can inform strategy/policy identification, assessment, implementation and evaluation include:

- **Identify potential worrying trends, assess the problem and its relation to the environment, analyse the causes of the issue, and how they impact society, the economy and the environment.**

In the first steps, we analyse the trends and identify the key causes of the issue to ensure that the problem is properly addressed and that the decision makers receive the identified information. This step extends the analysis to the impacts that the underperforming environmental trend may have on the other indicators.

- **Identify strategy/policy objectives and intervention options, and analyse the costs and benefits of each option.**

The identification of strategy/policy objectives is based on the outcomes of the issue identification phase and precedes the identification and choice of interventions. A cost–benefit analysis is necessary to identify the best and the poorest of the options in terms of the costs and benefits.

- **Measure the strategy/policy impacts across key sectors (business) and actors, as well as on society, the economy and the natural environment (well-being).**

The approach used to identify the strategy/policy impact indicators covers a broader set of consequences, which are of a social, economic and environmental nature. These indicators include information on the state of the environment, which is directly related to the environmental issues and target indicators, as well as the indicators of the sectoral performance and socioeconomic progress, such as employment and well-being.

### 3.2.2 Step 2: Identify the causes and effects

Once the problem has been delimited, key variables and actors directly linked to—i.e. affecting or impacted by—the problem need to be analysed. This further delimits the boundaries of the problem and facilitates creating influence tables and causal maps.

Rigorous logical steps need to be followed to identify causal relations: first of all, one has to ensure that the identified cause occurred prior to the assumed effect.

***An example:*** the causal relations between inflation and employment illustrate the possible pitfalls of this seemingly simple exercise. At first glance, one would assume that increasing inflation leads to more unemployment, since employers are forced to lay off employees in order to meet their costs. However, it is also possible that an increase in the work force could lead to more demand for goods, which in turn causes an increase in their prices. It is therefore very difficult to establish a causal relation without considering the specific context and the problem that needs to be solved.

Furthermore, the cause should clearly be directly linked to the effect. Although it may seem trivial, this step can be insidious. It should therefore be approached with utmost attention, since errors at this stage could jeopardise the entire problem-solving process. In order to carry out this task effectively, direct causes have to be distinguished from indirect ones.

As the previous example shows, the relation between unemployment and inflation is not a direct one. Inflation is directly linked with the company's capacity to meet its costs. It is thus indirectly linked with corporate employment policies. The same can be said for the opposite relation: the employment rate is directly linked with the demand for goods, which, together with other variables, ultimately determines the inflation rate.

The distinction between the direct (primary) and indirect (secondary) causes of a problem is relevant to find the main entry points for intervention and to evaluate how far-reaching the problem is.

***An example:*** the strategy that Toyota adopted in response to a crisis that it experienced between 1987 and 1991 is a good example of such a distinction being made. During these years, demand for cars was overheated, leading to a huge expansion of the car manufacturing industry. This sudden growth in demand led to a labour force shortage. Toyota managers carefully analysed the cause–effect relations in the system in order to respond effectively to this challenge. They identified a number of concurring causes of their inability to increase production and hire additional employees.

The primary cause that they identified was the decrease in new labour market entrants, which was due to the declining birth rates. The secondary causes included new workers' tendency to avoid heavy manufacturing jobs and the existing workers' reluctance to accept very demanding shifts on the assembly line. Since they couldn't address the primary cause and increase their staff, the managers decided to address the secondary causes. They made assembly work, and the workplace in general, more attractive to increase productivity and attract new talents. To achieve the latter two goals, the production efficiency objectives were made less stringent, more attention was paid to the rights of female and older workers, and the annual working hours were reduced (Shimizu, 2000). This strategy proved successful, allowing Toyota to hire more workers, thus gaining the competitive advantage required to respond to the growing demand for new cars.

Finally, a systemic approach implies the acknowledgement that a single effect can be the result of multiple causes and that a single cause can have multiple effects on the analysed system. It is essential to map and involve a wide range of stakeholders at this stage to obtain an all-round vision of the problem. This limits the risk of crucial causal links being ignored and ensures that key responsibilities within the system are identified.

***An example:*** a problem such as inadequate municipal waste management can have a variety of effects on different sectors and segments of the population. Lagos is a Nigerian megacity with more than 10 million inhabitants; its population growth, lack of civic culture, obsolete trucks, poor accountability, among others, have led to inadequate waste management. This has a variety of effects on the economy, society and environment: reduced agricultural production due to groundwater pollution from leachate, health problems due to ozone formation, fire hazards, an increase in disease vectors such as rats and insects from open landfills, and environmental degradation due to ozone formation and methane and carbon dioxide emissions. Under the coordination of the National Environmental Standards and Regulations Enforcement Agency (NESREA), different actors address these negative effects. At the federal level, for example, a number of ministries are involved. These include the Ministry of Water Resources—wastewater management; the Ministry of Environment—pollution from inadequate waste disposal and management, and climate change impacts on waste management; the Ministry of Energy—waste to energy technology; the Ministry of Health—prevention of diseases due to unsustainable waste management; and the Ministry of Lands and Urban Development—municipal waste management. Understanding the direct and indirect effects of inadequate waste management in Lagos helps delimit the problem and share responsibilities with a wide range of actors (Ogwueleka, 2009).

***An example:*** broadening participation in the development of corporate strategies can improve knowledge sharing across different company branches and levels, possibly leading to innovative solutions. For example, the US pharmaceutical firm Abbott has performed well in building a system that leverages internal knowledge. It has created a team with people from different departments within the firm to exploit new opportunities. Originating from a US$250 million HIV Aids prevention campaign in Africa and other developing countries, Abbott, together with the Gates Foundation, decided to pursue the idea of developing cures for neglected tropical diseases that were observed during the HIV Aids project. The Abbott Fund was created for this purpose. Front-line scientists, lawyers, regulators and fund representatives manage this fund. All the parties involved have contributed their distinct know-how and have applied their respective business protocols. In addition to its positive impacts on health in developing countries, the Abbott Fund has helped improve the company's global reputation (Zimmermann *et al.*, 2013).

### 3.2.3 Step 3: Analyse future behavioural paths and impacts

Once the root causes of the problem and their effects on the system have been identified and delimited, qualitative projections can be used to better understand the gravity and urgency of the issue and the relevance of each of the causes' key drivers.

On the basis of past and current trends and patterns, forecasts can be made regarding possible future developments of the problem. This will help identify these developments' ramifications across the system (e.g. across sectors and departments). The expectations should be sufficiently clear to be easily disseminated among all of the relevant stakeholders. Doing so will raise awareness of, and interest in, the problem, which will shape the debate.

> ***An example:*** in recent years, many companies have started incorporating climate change adaptation measures into their risk management plans, taking a multi-stakeholder approach to include public concerns in their planning exercises. The analysis of potential climate change impacts on the company activity involves short, medium and long-term projections based on systemic cause–effect analyses of worrying trends. For example, Anglian Water, a water and sewerage services company that operates in East Anglia, the driest region in the UK, has undertaken risk assessments to identify climate change adaptation priorities. Examples of key priorities include: increasing operational resilience to minimise impacts of extreme weather events (immediate); implementing preventive measures (e.g. water metering, water efficiency programmes and leakage control) to improve network resilience in the face of seasonal changes in climate (short to medium-term); investing in enhancing the climate resilience of current assets for the next 40 years (long-term) (Agrawala *et al.*, 2011). This is done to ensure high quality service, from the perspective of its clients, in spite of the increasing variability of the environmental conditions.

Exploratory storylines can be used to delineate the gravity of the expectations and the repercussions across the key stakeholders if no action is taken.

> ***An example:*** UNEP's Global Environment Outlook (GEO-5) emphasises the choices and strategies that could, from 2012 onward, lead to a sustainable future (UNEP, 2012a). UNEP does so by focusing on two very different storylines that are based on a review of existing scenario studies: 1) a view of the world in 2050 assuming

business-as-usual paths and behaviours—i.e. 'conventional world' scenarios—and 2) an alternative that leads to results consistent with our current understanding of sustainability, agreed-upon goals and the targets on the road to 2050—i.e. 'sustainable world' scenarios.

***Some examples:*** the GLOWA-JR project aims to provide scientific support to improve water management in the Jordan River region with particular emphasis on adapting to global climate change. The project consists of several phases with various objectives and outputs. Phase II built a framework to analyse the region-wide questions regarding global climate change and water resources. A scenario-building process was one of the main components. Four 'general development scenarios' were developed. They emphasised the effects of likely climate change and their impacts on the overall available water in the region. These scenarios were analysed in the context of possible alternative economic developments, as well as the overall willingness to move to a lasting peace in the region. The latter would imply sharing water resources (Centre for Environmental Systems Research, 2009).

Siemens has worked on identifying the megatrends that will shape our future. Scenario analysis is also carried out to evaluate the potential role of these trends—such as demographic change, urbanisation, globalisation and climate change—in shaping the company's corporate strategy. Internal and external factors are believed to increase Siemens' readiness to capitalise on future opportunities and adapt to changing market conditions (Siemens AG, 2013).

In 2008, AngloGold Ashanti, a South African mining company with global operations, commissioned a study to evaluate key climate risks and challenges that could destabilise its activities, as well as the well-being of the surrounding communities. The study highlighted a number of potential threats and worrying trends, such as increasing energy usage for cooling purposes, deterioration of the company's infrastructure due to increasing rainfall patterns, and the overall negative impacts of rising average temperatures on the well-being of employees and local communities (Agrawala *et al.*, 2011).

### Box 1 Proposed tools: indicators

Indicators, as the word suggests, are instruments that provide an indication, which is generally used to describe and/or give an order of magnitude to a given condition. Indicators are a crucial tool to identify and prioritise issues. They provide clear information on the historical and current state of the system, and highlight trends that can shed light on causality, which allows the key drivers of a problem to be better detected.

Within the integrated decision-making process, indicators can be used to: 1) identify issues and their primary causes; 2) carry out a cost–benefit analysis to evaluate intervention options; and 3) support integrated monitoring and evaluate strategy/policy impacts (UNEP, 2012b).

Indicators can be divided into three main categories:

- **Problem identification indicators.** These indicators seek to facilitate the identification of issues. Regardless of the nature of the problem to be solved (environmental, social or economic), indicators are selected that best identify the problem and its, at times many and varied, causes and effects
- **Strategy/policy formulation and assessment indicators.** This group of indicators assesses the potential costs and performance of various intervention options that could be utilised to solve the issue
- **Strategy/policy evaluation indicators.** These indicators aim to assess the success of strategy/policy interventions. Impacts have to be calculated by means of an integrated approach to strategy/policy evaluation, which includes the progress of: 1) human well-being, especially with regard to public policies; and 2) other operations in the business if the private sector is involved

| Strengths | Weaknesses |
| --- | --- |
| • Support problem identification<br>• Allow problems to be quantified and their causes and effects to be identified<br>• Facilitate the objective evaluation of intervention options<br>• Enable quantitative monitoring and the evaluation of implemented actions | • Data is not always reliable and coherent across sectors<br>• Data needs to be combined with other tools for a systemic analysis (e.g. causal maps, scenarios, simulations, etc.) in order to effectively support decision making |

TABLE 2 Strengths and weaknesses of indicators

Indicators support all steps of the decision-making process. They therefore complement the other tools in this book well.

The approach proposed consists of utilising indicators to identify the problem. Influence tables and causal loop diagrams (CLDs) allow for a more systemic analysis of causal relations to identify the key causes and effects of the problem to be solved. Indicators are then used to set targets and identify possible strategy/policy options to mobilise the required investment. CLDs allow for testing the effectiveness of interventions by identifying their key entry point, as well as their direct and indirect impacts throughout the system. Finally, indicators are used to select key variables to create scenarios.

## Case study 1 Problem identification with systems thinking: the creation of the Ethiopia Commodity Exchange

The creation and operation of the **Ethiopia Commodity Exchange (ECX)** can provide relevant insights into a successful approach to problem identification based on the understanding of the functioning mechanisms of the system and the roles of various stakeholders (Stadtler and Probst, 2013).

Dr Eleni Gabre-Madhin, former CEO of the ECX, discovered that starvation in many parts of the country was due to the ineffective trading system, which failed to deliver food to the remote and most vulnerable areas. In order to better analyse and understand the problem, she conducted a national grain and coffee market survey spanning 45 markets and hundreds of traders, and she travelled around the country following the grain from where it was produced to where it was sold. In particular, she realised how inefficient the Ethiopian trading system was after conducting an in-depth interview with Abdu, a local trader who had lost considerable amounts of money trying to deliver maize to the northern part of the country. His 900 km trip was delayed due to road checkpoints and poor road infrastructure: instead of the planned three days, it took him two weeks to arrive at his destination. When he finally arrived, the buyer refused to seal the deal because the maize quality was low, and prices had gone down meanwhile. As a result, Abdu was obliged to sell at a loss and return home.

Based on these observations, Eleni stressed the need for an institution that would help facilitate all the crucial phases of food exchange by: 1) introducing standards; 2) grading quality; 3) coordinating trading; 4) developing reliable payment and delivery systems; and 5) ensuring contract enforcement, among others (Stadtler and Probst, 2013). This led to the creation of the ECX, a commodity exchange designed to be a holistic platform that would integrate all of the elements mentioned above. Additionally, the ECX was created as a multi-stakeholder entity, representing the interests of all actors, across the public and private sector.

ECX is led by a board composed nearly equally of public and business representatives, responsible for governing and controlling the exchange. As stressed by Eleni:

> The public sector brings the social interests to the table, keeping an eye on the question: Is this in line with the overall development interests of the country? The private sector, in turn, emphasises business elements, such as efficiency and cost-effectiveness (Stadtler and Probst, 2013).

Thanks to this integrated approach, the trading volume went from 138,000 metric tons in the first year to 222,000 tons by the second (mostly coffee), and to 508,000 tons in the third year. The ECX has become a financially sound company, running its operations on a tight basis, with minimal fees charged to the market.

To conclude, this case study shows how problem identification is an essential phase of the strategy/policy cycle. Eleni was able to adopt a systemic approach to problem identification: she started from the observation of events (e.g. Abdu's failed deal) and undertook a deeper analysis of the underlying causes of the problem by exploring patterns (i.e. chronic malnutrition in some areas of the country), systemic structures (e.g. absence of standards and trade regulations) and mental models (i.e. traders used to rely exclusively on word of mouth to know whether a person was trustworthy) influencing the behaviour of the system. Based on this analysis, Eleni identified the main problem as the absence of an integrated approach to grain trade in Ethiopia: the different elements composing the production and distribution chain were not effectively connected. Her work led Ethiopia away from potentially devastating socioeconomic impacts.

# 4
# Phase 2
## System characterisation

**Conceptual mistake #2a:** Every problem is a direct consequence of a single cause.
**Conceptual mistake #2b:** We only need an accurate 'snapshot' of the actual state of the system to find solutions.

**What to do:** Map the complexity and explore the dynamic properties of the system.

**Steps:**
1. Build a causal diagram and review the boundaries of the system
2. Create a shared understanding of the functioning of the system
3. Identify key feedback loops and entry points for intervention (strategy/ policy identification)

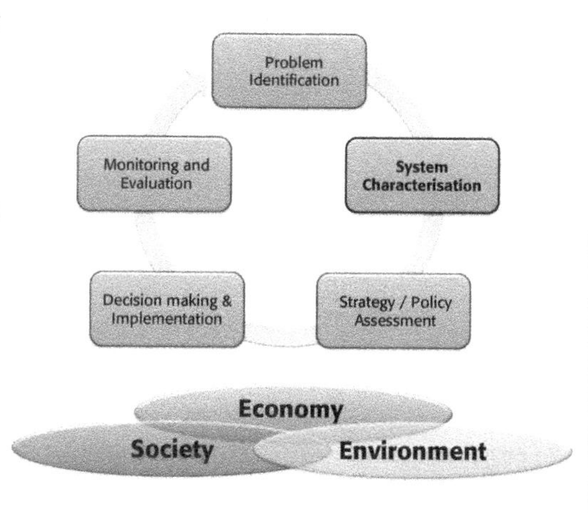


**Tools:**
- Indicators
- Influence tables
- Causal loop diagrams
- Scenarios

## 4.1 Conceptual mistake #2a: Every problem is a direct consequence of a single cause

A common mistake when dealing with the real world's overwhelming complexity is using extremely simple causal models to solve complex problems (Dorner, 1980; Einhorn & Hogarth, 1986; Damer, 2009). A single cause is then often identified for any given problem, which leads to erroneous analyses and partial solutions. The underlying assumption is that a single effect (or event) is attributed to a single cause and if that one cause is addressed, this will solve the problem. This approach leads to an extreme simplification of the real world. Consequently, it greatly reduces the effort of having to explore the system to identify concurring causes, multiple interconnections, feedbacks and other elements of dynamic complexity.

Failing to identify multiple causes leading to a single problem is often the result of 'siloed' approaches to problem solving. Complex problems tend to cross boundaries and affect different economic sectors, social groups and environmental issues across the planet. Nevertheless, most decision makers still focus on their specific area of expertise. They often ignore the parallel decisions that their counterparts, who operate in related fields, have made and blame unwelcome change on others and/or on external conditions that cannot be influenced.

In conclusion: without continuous and effective communication between decision makers across sectors and branches, the environment's dynamic complexity and the importance of feedbacks across sectors are likely to be overlooked.

***An example:*** many different causes can be identified and addressed to solve a complex problem such as the availability of potable water. However, a siloed approach might miss the potential synergies between public policies and private sector strategies. This greatly increases the risk of implementing conflicting measures. For example, in 2012, China announced a US$372 billion plan for energy efficiency and pollution control, with a specific focus on water pollution from industrial waste. However, the Chinese government simultaneously subsidises the national industries that produce chemical fertilisers for agriculture.

They thus artificially lower the price of substances that greatly pollute groundwater (Huang *et al.*, 2012). Owing to these two conflicting policies, public expenses and private investment have both increased, but the total amount of potable water is likely to remain unchanged.

## 4.2 Conceptual mistake #2b: We only need an accurate 'snapshot' of the actual state of the system to find solutions

Ignoring the dynamic interrelations between the multiple variables in a system can lead to another common mistake in problem solving: analysing a static picture of the system (Simon, 1982; Brehmer, 1992). This approach is based on the underestimation of time as a factor of change. In particular, the relation between the history of a system (stocks) and its current, continuous modifications (flows) tends to be disregarded, or considered of secondary importance. This situation is therefore implicitly considered motionless and unalterable if no decision is taken.

***An example:*** the tourism business in small island developing states (SIDS) often considers only the current state of its natural resource stocks (e.g. coral reef, fish variety, beaches). This approach to tourism development is only profit-oriented. Further, it is based on the assumption that nature will always provide the ecosystems that make islands attractive travel destinations. It ignores tourist development's unsustainable impacts on the health (and value) of the ecosystems on which the industry relies. This business model thus leads to increasing environmental problems such as coral reef degradation and coastal erosion.

Aggressive tourism policies—such as the expansion of the hotel capacity—cause or aggravate these problems. They are addressed by relocating high-quality hotels and resorts (or their brands) to virgin areas, where the ecosystems have not been compromised. In the short term, this model pays off because tourist activities are kept high by lowering the price of services in the degraded areas and increasing it the new areas. However, this is an unsustainable approach with disastrous consequences for the local economy, society and environment in the medium and long term.

Furthermore, tourism has also contributed to the clearance of coastal mangroves in, for example, many Caribbean islands. In Negril, Jamaica, up to 55 metres of beach depth has been lost as the result of coral reef degradation caused by unsustainable tourism activities and climate change impacts. Hotels and resorts are still making considerable profits, but degradation of the coastal ecosystems has increased the storm surge risk. In the long term, increased extreme weather events are likely to affect more than 60 hotels, their guests, and the water and sanitation infrastructure (UNEP, 2010b).

An analysis of the 'snapshot' of the actual state of the system thus fails to consider crucial information about how the variables underlying the system change over time. If the evolution of the system's various components that lead to the problem is ignored, this prevents decision makers from understanding the true causes of change and adequately forecasting future changes (e.g. how the system would respond to implemented interventions). A missing or incomplete analysis of trends is therefore a critical limitation of the decision-making process.

***Examples:*** Narragansett Bay (USA) has experienced fish kill or die-off a few times in the last decade. An analysis of the present state (or of a snapshot) of the system biology cannot identify the actual causes of this large-scale mortality. Although hypoxia—the lack of oxygen in the water—is the immediate cause of the fish dying, this is the result of the interaction of several conditions. Unfortunately, these conditions are often not analysed simultaneously, nor is their evolution over time. An understanding of how the system functions is crucial to identify and implement effective interventions.

A company could decide to invest in the production and distribution of a drug because of the high incidence of a given disease. However, a dynamic analysis of the market would have revealed that other companies had already started developing the same drug a year earlier. Because the process takes 48 months, the company would lag far behind its competitors.

Neglecting feedback loops is another consequence of a static analysis of complex, dynamic systems. This prevents potential risks and opportunities from being identified, which are derived from the reinforcing or balancing character of feedbacks within a system. A dynamic understanding of a problem helps identify the variables. These entry points for policy interventions could be targeted to amplify (or balance) a given trend. Similarly, the potentially threatening effects from feedback loops (i.e. strategy/policy responses) could be detected in advance and used to guide the decision-making process.

> ***Examples:*** based on a static 'snapshot' of the market, managers of a postal company might notice that the level of shipping requests is well below the corporate goals. The managers' immediate response might be to downsize the personnel in order to reduce the costs and gain competitiveness. However, the shipping requests may be declining due to consumer dissatisfaction with delivery delays.
>
> Low agricultural production in a country might encourage the policy-makers to support the use of soluble salt fertilisers, which will increase the soil fertility and crop yields. However, soluble salt fertilisers reduce the biological health of the soil in the medium and long term, thus leading to a progressive decline in agricultural production.

## 4.3 What to do: Map the complexity and explore the dynamic properties of the system

To address the methodological shortcomings illustrated above, the complexity of the system needs to be analysed and appreciated. This allows one to identify and isolate its different components, to highlight the causal linkages between them, and to understand the dynamic relations that determine the behaviour of the system as a whole. A systemic approach to problem analysis is likely to reveal the multiple concurring causes of a problem and their varying effects on various parts of the system. Identifying the causal relations and understanding how they influence a problem, will allow one to design more targeted and effective interventions. These interventions will fully consider the dynamic properties and complexity of the system. However, one should remember that all these different causes and effects are part of a single system. If one variable changes, all the others—including the causes and effects of a problem—will also change. From this point of view, a system lives, grows and learns over time.

> ***An example:*** many countries face a common public policy problem–their energy demand exceeds their energy supply. A systemic analysis will reveal that the population, economic activities, energy prices and energy efficiency (technology) are the primary drivers of energy demand. On the other hand, labour and capital, which investments also affect, drive energy supply. Demand and the availability of funding–primarily a function of economic performance–generally impact the level of investment. This indicates that there is a feedback loop or a circular relation in the system. Consequently, any action aimed at influencing the demand for, or supply of, energy will ultimately affect the other actions as well.

> ***An example:*** it is essential that manufacturing companies map the variables influencing every step of their production and distribution chain. This will allow them to evaluate the potential occurrences of deficiencies in the quality of their product and delays in its delivery, which will also reveal their vulnerability to disruptions in the transport process.
>
> Among others, the dairy supply chain involves aspects related to inputs such as farm development, maintenance of the infrastructure, transport modalities and consumer habits. All these elements should be included in the 'map' of the dairy production and distribution system as they all impact the price formation at the farm, wholesale and retail levels. In addition, the effect should be analysed that urbanisation and policy changes, or institutional and technological changes, have on the dairy industry. In South Africa, for example, milk consumers were congregated in the inland provinces, while milk production was concentrated along the coast. This led to shortages in the inland urban areas. The distance between the producers and the consumers forced the dairy industry to transport

milk, which required gasoline and refrigeration, creating extra costs. Moreover, stricter milk hygiene regulations were introduced, which meant additional costs for producers who did not comply with the standards. These costs also included purchasing advanced technology to maintain the cold chain economically (South African Food Pricing Monitoring Committee - FPCM, 2003). Understanding these complex cause–effect relations is crucial if producers wish to increase their competitiveness and market growth.

A three-step approach is proposed to map the complexity of a system. In turn, this will allow winning strategies and policies to be formulated:

1. Build a causal diagram and define the boundaries of the system
2. Create a shared understanding of the functioning of the system
3. Identify key feedback loops and entry points for action

### 4.3.1 Step 1: Build a causal diagram and review the boundaries of the system

In the problem identification phase, problem boundaries are carefully defined by identifying the multiple cause–effect relations between the variables that the problem influences, or that are influenced by it. This previous understanding of the key system variables and their interactions allows a map of the system to be created. This is an essential step in the integrated decision-making cycle. In this cycle, an issue is analysed—taking the underlying complex network of variables and feedback loops into consideration—to identify key strategic entry points for intervention. While an intervention can have very positive impacts on certain elements of the system, it can create issues for others. An intervention may thus generate synergies and/or bottlenecks. Furthermore, successful longer-term interventions may have negative short-term impacts, but mitigating actions can be implemented to counter these.

***An example:*** controlled deforestation may initially seem a good decision for the private and public sectors. However, deforestation reduces sediment retention and increases the likelihood of flash floods, which increase siltation. Over time, this problem may reduce the use of rivers for timber transport, increasing the costs for the private sector, and requiring dredging, for which the government often has to pay. Mapping the complexity of the system can support the identification and mitigation of these side-effects, which happened in Borneo (Van Paddenburg *et al.*, 2012).

Causal loop diagrams (CLD) provide a comprehensive understanding of the cause–effect relations between the different variables contributing to, or being affected by, the problem. CLDs can therefore also help decision makers find entry points for intervention and test the potential effectiveness of the designed strategy.

**How to read a causal diagram**

Causal loop diagrams include variables and arrows (called causal links), with the latter linking the variables together with a sign (either + or –) on each link indicating a positive or negative causal relation:

- A causal link from variable A to variable B is positive if a change in A produces a change in B in the same direction
- A causal link from variable A to variable B is negative if a change in A produces a change in B in the opposite direction

These causal relations are presented in the following table:

| Variable A | Variable B | Sign |
|---|---|---|
| ↑ | ↑ | + |
| ↓ | ↓ | + |
| ↑ | ↓ | – |
| ↓ | ↑ | – |

Circular causal relations between variables form causal, or feedback, loops. 'Feedback is a process whereby an initial cause ripples through a chain of causation ultimately to re-affect itself' (Roberts *et al.*, 1983). These can be positive (amplifying change, and identified by a 'R' notation, for reinforcing) or negative (countering and reducing change, and identified by a 'B' notation, for balancing).

***An example:*** a decline in fish catches and in the size of the fish caught is a common problem that coastal countries have faced in recent years. The Seychelles Fishing Authority, for instance, recorded a 16% decline in the number of tuna fish caught in 2011 compared with 2010 (Seychelles Fishing Authority - SFA, 2012). The lack of mature fish can be due to a number of possible causes that influence the natural regeneration process. Figure 5 is a causal map showing the relations between some of the variables that influence the fish regeneration system. The natural feedback loop of fish regeneration consists of the balance between fish births and deaths. There is an increasing trend towards catching young fish, which prevents the fish from becoming mature. This creates a negative feedback loop that destabilises the system and makes a return to previous levels of fish catches increasingly unlikely. As the causal links in the diagram show, the number of mature fish depends on the relation between the actual and desired mature fish. Fishermen would catch less young fish if the mature fish they desire were to increase—they would even be willing to wait until these fish are mature. However, catches also depend on the fishermen's perception of when fish are mature—which depends on their experiences, the local culture, their mental models, etc.—and the specific daily catch required.

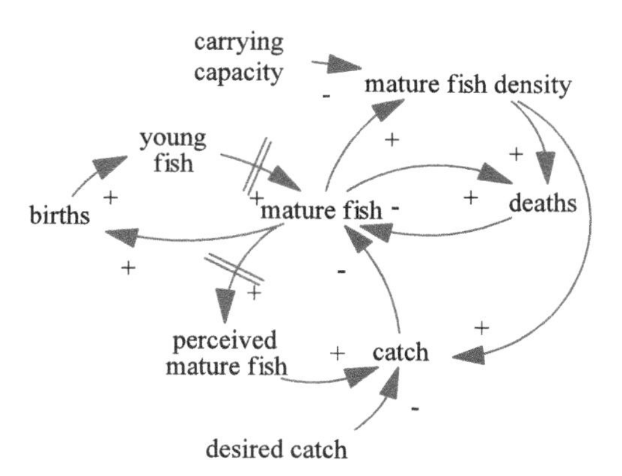


**FIGURE 5 Causal map of the fish stock regeneration system**[2]

2 More information on causal loop diagrams is provided in the Annex, Section A1.3.

In order to build a causal loop diagram, a series of logical steps should be followed. The variable most closely related to the problem to be solved—let's call it the 'central variable'—is the first item that should be added to the diagram. Thereafter, the variables directly linked to the problem ('first variables') have to be added by means of causal links, which are shown as arrows connecting two variables. Subsequently, the variables affecting the 'first variables' are added, etc. The process follows this structure until decision makers believe enough variables have been introduced to represent the complexity of the system.

Causal loop diagrams also help identify possible cascading effects that may characterise the problem positively and negatively. The impacts of a certain issue can cause other problems across the system, exacerbating its overall performance. Conversely, businesses look for the 'engine' drivers of growth. In this context a business model needs to strengthen positive, reinforcing loops while curbing balancing ones. Therefore, both the causes of problems and the impacts of interventions need to be carefully examined by means of a systemic perspective before the interventions are decided and implemented.

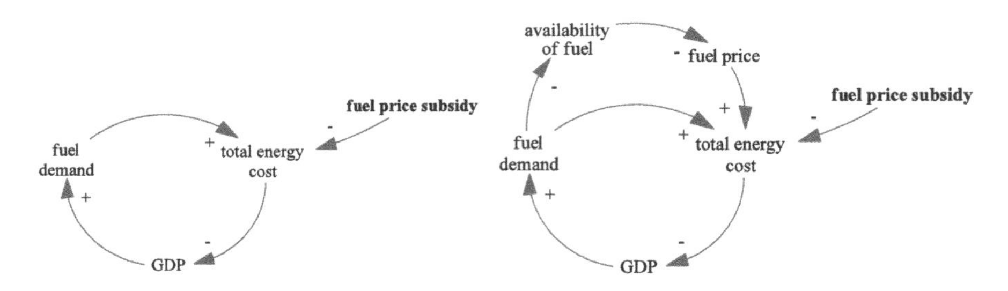


FIGURE 6 Simple causal diagram representing the main variables influenced by the introduction of fuel subsidies

***An example:*** the introduction of energy subsidies impacts energy costs—or households and companies' energy bills—although they do not directly affect the demand or consumption. If the energy costs are decreased, subsidies tend to increase the profitability, and the GDP, which is one of the primary factors that drive demand. If all else remains equal, higher demand increases the energy costs. Further, higher demand may decrease the domestic availability of fuel, which requires an increase in production or higher (and more expensive) imports. If this is the case, fuel prices and, therefore, demand are also likely to increase in the medium term. In spite of a decline in the energy costs in the short term, both of these factors will increase the energy costs in the medium and longer term, possibly offsetting the initial gains. This example is presented in more detail at the end of the chapter.

**How to create a causal diagram?**

The basic knowledge needed to build a CLD includes the concept of polarity (i.e. the sign of the causal relation between two variables, whether positive or negative) and the concept of feedback (reinforcing or balancing), as mentioned above. The following are the practical steps that should be followed:

- Start with the key indicator identified as representing the problem and add it to your diagram (which is blank at this stage)
- Add the causes of the problem, one by one, linking them to the first variable considered and determine the polarity of the causal relation
- Continue identifying and adding the cause of the cause, and so forth

In the process, the diagram will grow and other variables will influence some of the variables identified as causes of the problem. These circular relations are the feedback loops (representing closed-loop thinking), which are also the key functioning mechanisms of the analysed system. Thinking in terms of feedbacks is crucial in the development of CLDs and requires a multi-stakeholder approach.

More specifically, the following recommendations should be followed to create a good causal diagram:

- Use nouns or noun phrases to represent the elements rather than verbs. For example, use 'cost' and not 'increasing cost' as an element. Use a variable name in a positive sense. For example, use 'growth' rather than 'contraction'
- There are often differences between short-term and long-term consequences of actions and these may need to be distinguished with different loops
- Keep the diagram as simple as possible, subject to the earlier points. The purpose of the diagram is not to describe every detail of the management process, or the system, but to show those aspects of the feedback structure that lead to the observed problem. In other words: model the problem, not the system

In essence, on the one hand, mapping the complexity of a system avoids excessive simplification of the reality in which a problem takes place. On the other hand, creating causal diagrams helps decision makers focus on the most important variables, relations and feedbacks in order to formulate appropriate, context-specific interventions. By breaking the system down into its elements, causal diagrams help visualise and provide a better understanding of the problem's complexity, and guide decision makers effectively through the problem-solving process.

### 4.3.2 Step 2: Create a shared understanding of the functioning of the system

If a wide variety of stakeholders participates in creating a causal loop diagram (e.g. group model building as a part of scenario analysis), this is likely to maximise the results in terms of the quality and ownership of the process and the diagram.

Creating a map of the system has the potential to bring the ideas, knowledge and opinions of multiple relevant stakeholders together. This ensures that different points of view—as well as potential causes—are considered simultaneously and objectively. In turn, this allows all the stakeholders to gain basic to advanced knowledge of the systemic issues. They can then collectively design effective intervention options in a true multi-stakeholder process.

Once the causal loop diagram is finalised, the contributors will recognise their personal and collective knowledge in the map. Subsequently, they can explore the dynamic interplay of the key variables in shaping the system's historical and present behaviour. A discussion of each variable and causal relation's relevance can then be initiated to ensure that the causal loop diagram represents reality objectively. In a participatory effort, some of the links may be removed during the discussion, while new ones are added. In general, the discussion of the linkages is likely to create an environment conducive to improved collaboration between the relevant stakeholders—each participant can visually identify the multiple links between his or her work and that of the other stakeholders.

***An example:*** Maui, Hawaii, which is often described as paradise on Earth, is one of the most desirable tourist destinations in the world. But for those who call Maui home, their paradise is slowly turning into a grim reality. Despite the island's aesthetic appeal and tropical climate, the resident community on Maui, about 185,000 people, faces serious problems with allocating and managing its diminishing water supply. Over the years, companies developed ditch, tunnel and reservoir systems to redirect water in order to satisfy the growing demand on the leeward side of the island. Considered an engineering marvel of its time, the diversion system has grown to the point that almost 274 million gallons per day of surface water are now diverted each year, with almost half of that water originating from the eastern part of the island.

Given that the water system on Maui also holds a strong cultural relevance and value for the local population, a multi-stakeholder approach was employed to map the water system, as well as all the social, economic and environmental driving factors of demand and supply. Group-modelling sessions were held with the local stakeholders representing the farmers, landowners and other concerned citizens. Through this process, local needs and concerns were explored and included in the causal loop diagram of the water model that was later developed. The group session allowed for including possible policies and actions in the diagram. These were discussed and evaluated with local experts to better assess the role and relevance of the local cultural and structural constraints. Further, despite the initial disparity in opinions on which data to trust and which action would have the best outcome, creating the diagram allowed all the participants to visualise the key causes and effects. They could then agree on the main action items that they needed to investigate further (Bassi & Mistry, 2009; Bassi *et al.*, 2009b).

If the perception that individuals or groups are part of the broader picture is strengthened, they tend to understand the cascading effects of their actions, whether positive or negative. Such an understanding leads to enhanced accountability, openness to dialogue and willingness to work together towards common solutions to complex problems.

***Examples:*** the incentive package for renewable energy production in China is an example of the virtuous combination of different actions to solve a problem. Command-and-control as well as market-based regulations were adopted by involving different stakeholders in the policy formulation. This package incentivises small and large producers to invest, as it includes a feed-in tariff, a national fund to support investment and tax reductions for renewable energy projects.

Another example is the Loess Plateau revitalisation: home to more than 50 million people, the Loess Plateau in China's northwest has been subject to centuries of overuse and overgrazing. This led to very high erosion rates and widespread poverty. The multiple causes of these phenomena were identified: they included the high population density, unsustainable agricultural production and deforestation, among others. This paved the way to implement two large-scale public works projects to counter the causes. These projects sought to rehabilitate damaged and degraded ecosystems by means of multiple interventions, which included tree planting, terracing, restrictions on grazing, constructing reservoirs, soil enhancement and conservation techniques. Previously, single policies failed to obtain the expected results. In fact, earlier limitations imposed on grazing impacted the livelihoods of the local communities. They therefore partially replaced herding with extensive agriculture, which led to the logging of trees in order to expand the available arable land. In order to contain this excessive extension of agricultural land and prevent deforestation, the authorities decided to enhance the land's productivity by adopting targeted interventions to improve the soil fertility and efficient use of water. Through the enhanced land productivity project, certain areas could be set aside to plant trees and restore the forest ecosystems. These in turn helped reduce soil erosion. Each of these solutions, if implemented in isolation, would have worsened the problem instead of solving it.

Private companies are also increasingly using participatory approaches to create new ideas and increase their potential for innovation. In particular, collaborative idea management is a strategy adopted to involve employees in the resolution of complex problems and the identification of new ideas for the company's development, thereby improving communication and the shared understanding of the challenges and opportunities. IBM, for instance, has created the ThinkPlace programme, a platform where anyone from within the company can make suggestions, share problems and ask for advice. This collaborative approach to problem solving allows the managers to gain a more comprehensive understanding of the dynamics governing the system, and to identify solutions and new ideas based on a variety of different inputs and perspectives.

### 4.3.3 Step 3: Identify key feedback loops and entry points for action

The creation of a map of the system and the shared understanding of the causes and effects of the problem are essential to identify possible entry points for action. However, creating causal loop diagrams is not sufficient to effectively inform decision making. The functioning of the system needs to be investigated and the factors that trigger change—including the positive and negative feedback loops—have to be analysed. Feedback can be defined as 'a process whereby an initial cause ripples through a chain of causation ultimately to re-affect itself' (Roberts *et al.*, 1983). While the causal relations provide a sense of direction, two or more variables—through a rather static representation of the system—provide feedbacks that highlight the strength and dominance of certain parts of the system. These are the parts that generate change. In essence, feedback loops highlight the complex and dynamic structure of the system.

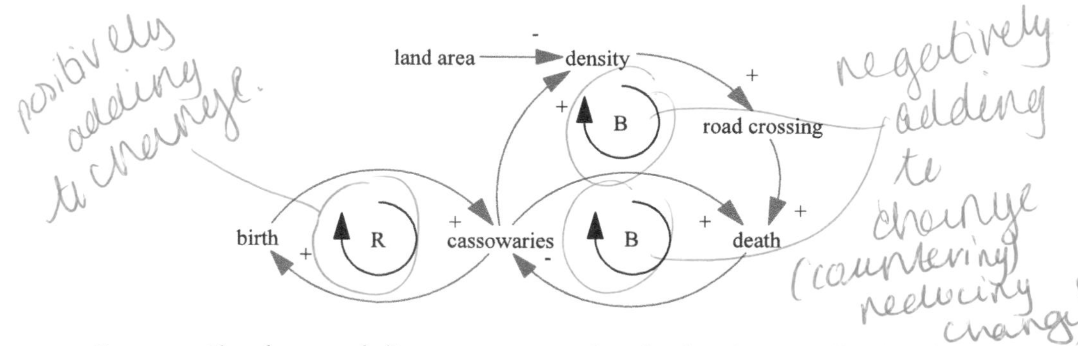


FIGURE 7 Simple causal diagram representing the key factors affecting the cassowary population near Cairns

***An example:*** the cassowary population is very important in Australia, especially in the northeast. Cassowaries roam in forested areas and generally feed on fruit. Cyclone Yasi damaged most of the forest in the area in February 2011 and the cassowaries were left with little food. In their search for food, they started crossing roads more frequently to reach new, less damaged areas of the forest, which increased their mortality further. Fenced-off areas were not the answer. This would probably have increased the mortality even more as the cassowaries would have been confined to land with no fruit. In general, this situation sheds light on the key feedbacks dominating the health of the cassowary population.

Crossing the road is not required if there is adequate food for the entire population, but it becomes more relevant when food is scarce, i.e. when there is an overpopulation of cassowaries, or when their feeding ground is suddenly reduced. It is no surprise then that equilibrium has to be reached—by balancing the feedback loops—to ensure the right balance between the cassowary population and the food available for them. The road crossing is an added threat and a balancing factor. The best intervention options were to provide fruit in the very short term—preventing the cassowaries from starving—and to carefully manage the land and ecosystems in the medium and longer term—increasing the ecosystem goods, or fruits, thus supporting the cassowary population better.[3]

***An example:*** obesity occurs when energy intake from food and drink consumption is greater than energy expenditure through the body's metabolism and physical activity over a prolonged period, resulting in the accumulation of excess body fat. However, studies indicate that there are many complex behavioural and societal factors that combine to contribute to the causes of obesity. Applying systems thinking to this problem allowed a 'complex web of societal and biological factors that have, in recent decades, exposed our inherent human vulnerability to weight gain' (Butland *et al.*, 2007)

3 Excerpt from a System Dynamics course offered by Dr Bassi at James Cook University, Cairns, Australia, 2011.

to be identified. An obesity system map was created by the Foresight team of the UK Government, including the energy balance at its core. Three main feedback loops were identified in this core diagram, including two balancing and one reinforcing.[4]

Identifying feedback loops allows one to isolate the main reinforcing and balancing forces in a system. Moreover, feedback loops highlight the potential creation of side-effects and synergies across the variables and sectors. Two types of feedback loops interact within a complex system and can be identified in a causal diagram: positive (self-reinforcing) loops, which reinforce or amplify the existing trend, and negative (self-correcting) loops, which tend to maintain (or reach) equilibrium. The values of individual variables often have a tendency to oscillate within a given range.

Positive loops are generally marked in the causal loop diagram with the letter R (reinforcing) inside a circular arrow; negative loops have the letter B (balancing).

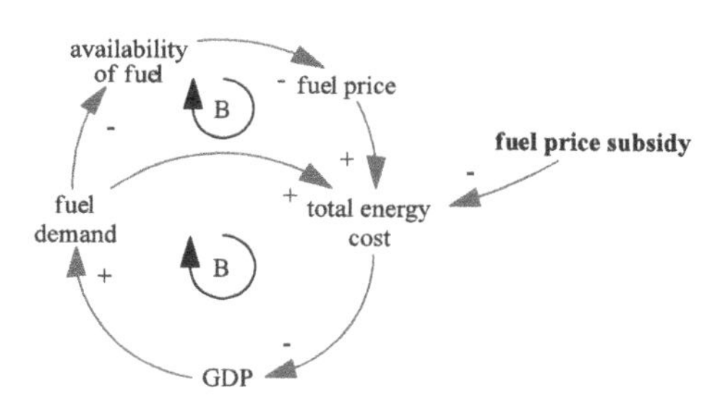


**FIGURE 8 Simple causal diagram representing the main feedback loops influenced by the introduction of fuel subsidies**

If subsidies are introduced, two balancing feedback loops can be identified in the simple causal loop diagram. These indicate that fuel subsidies can be effective in increasing the GDP in the short term, but ultimately the system tends to balance—unless other interventions are implemented. Consequently, when the role of subsidies is considered in isolation, it is obvious that this intervention will not have lasting positive impacts on the GDP.

4 More information on the system map can be found at www.bis.gov.uk/foresight (accessed 18 October 2013).

Further, the complex dynamic interactions between the positive and negative loops need to be explored when evaluating any strategy/policy. More specifically, the feedback dominance has to be estimated. The behaviour of the system demonstrates the existence and dominance of the feedback loops in the system. According to Richardson and Pugh (1981), 'a loop that is primarily responsible for model behaviour over some time interval is known as a dominant loop'. It is essential to identify dominant loops in a system in order to select entry points for intervention. On the other hand, linear mental models often lead to a misperception of the feedback dynamics. This may undermine the entire problem-solving process and inevitably lead to failure.

**How to identify the strength of causal relations**

Influence tables are useful to identify the causal relations among key variables in the system, as well as the strength of these relations. This tool allows decision makers to visually identify the key drivers of change, as well as the key effects of changes in the key variables in the system.

In order to facilitate identifying the strength of the causal relations and entry points for action, the **active sum** (AS) and **passive sum** (PS) are calculated. A **quotient** (Q = AS/PS*100) and a **product** (P = AS*PS) are also estimated for each variable. These are used to characterise the variables as follows:

- **Active variable** (highest Q). Influences the others a lot, and is little influenced by others
- **Passive variable** (lowest Q). Influences the others very little, and is a much influenced variable
- **Critical variable** (highest P). Influences other variables strongly, and it is also strongly impacted by other variables in the system
- **Inert variable** (lowest P). Has a limited influence on other variables, and is only slightly influenced by them

Characterising key variables as active, passive, critical or inert provides a good starting point for decision makers to identify the main levers for intervention. When coupled with the delay of each specific intervention option (concerning implementation and/or system response), influence tables effectively inform the creation of causal loop diagrams and decision making.

In the context of decision making, identifying the key feedback loops in the causal diagram allows one to detect the forces that should be strengthened and weakened through strategy/policy intervention. This way, it is possible to design an intervention that uses existing feedback loops and avoids the creation of potential side-effects.

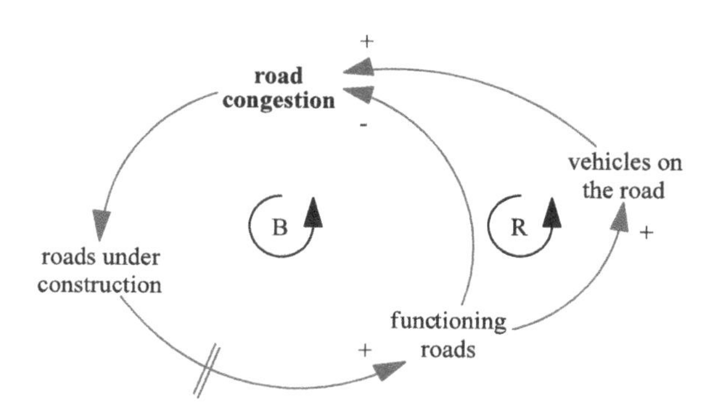


**FIGURE 9 A simplified representation of a road congestion problem. The crossed arrow represents the existence of delays between the beginning of road construction and road availability**

The simplified causal diagram shown in Figure 9 indicates that two feedback loops impact road congestion. The balancing one (B) indicates that if the road network is expanded, congestion will decline. On the other hand, the reinforcing loop (R) indicates that if more roads were available, this would lead to more vehicles on the road. Consequently, strengthening the balancing loop would not reduce the road congestion, as the reinforcing loop would counter any intervention. A more lasting solution would be to focus on reducing the strength of the reinforcing loop by introducing other means of transport, or by lowering the attractiveness of road transport (e.g. through taxation, which London, Milan, Bergen and several other cities in Europe have implemented).

Finally, only a few selected variables in the system can be directly influenced through external intervention (i.e. strategy/policy). As a consequence, it is important to identify which variables in the causal diagram decision makers can influence. This allows interventions, which identify and evaluate specific options that influence the system, to be targeted and to estimate their indirect and induced effects across all the key variables identified in the causal diagram.

> ***An example:*** worker motivation is crucial to ensure that work is done on time and the quality is high. On the other hand, despite motivation's crucial role in determining business success, decision makers cannot influence it directly. However, they can affect it indirectly through, among others, a salary system and/or the adoption of flexible work schedules. On the other hand, it is not possible to confidently forecast how these actions will impact motivation.

Such evaluation should also be done by using scenario analysis, which allows decision makers to assess the effectiveness of their interventions with varying assumptions about external events and trends. This additional layer, presented in more detail in the decision-making phase, adds risk assessment to the strategy/policy formulation process.

## Box 2 Proposed tools: influence tables

An influence table is a matrix that relates every (selected) variable of the system to the others and asks how strong the causal link (or influence) between these variables is.

The strength is generally defined using a scale from 0 (no influence) to 3 or 5 (strong influence). The blanks along the diagonal of the matrix show that the individual variables cannot influence themselves directly.

Influence tables are useful to identify the causal relations among the key variables in the system, as well as the strength of these relations. This tool allows decision makers to visually identify the key drivers of change, as well as the key effects of changes in the key variables in the system. They also provide a good starting point for decision makers to identify the main entry points for intervention.

Influence tables primarily support the decision-making process in its early phases. This tool can be used in the problem identification phase, during which causal relations are identified and their strength is assessed. Further, influence tables support the monitoring and evaluation stage, during which system-wide impacts can be assessed.

| Strengths | Weaknesses |
|---|---|
| • Support the identification of causal relations and their strength<br>• Identify key drivers of change in the system<br>• Allow entry points for action to emerge from the analysis<br>• Intuitive to use, straightforward interpretation of results<br>• Favour multi-stakeholder engagement in strategy/policy formulation | • Only present relations between two variables at a time. Multiple influence can only be inferred<br>• Analysis is static<br>• Cannot fully support strategy/policy formulation and assessment; needs coupling with causal diagrams |

TABLE 3 Strengths and weaknesses of influence tables

Influence tables can be used in synergy with indicators and causal loop diagrams. Influence tables are an intuitive and solid intermediary step between them, which greatly enhances the understanding of the issue/opportunity and cooperation among team members and stakeholders.

Influence tables are, with causal diagrams, a very important tool used to ‘see’ and explore systems in the decision-making process. They do so using a matrix format, with which most decision makers would normally be familiar (as opposed to the learning curve for the correct creation and use of causal diagrams).

• **For an example of an influence table, see page 143**

## Case study 2 Representing and analysing dynamic complexity to design effective policies: fossil fuel subsidy reform

A useful example to deepen the analysis of the system characterisation phase is fossil fuels subsidy reform (IISD, 2013).

In recent years, many governments have committed to support the shift towards more sustainable production and consumption modes through the implementation of policies and the allocation of investments that foster the development of green sectors. Among the key policy instruments that might create the enabling conditions for a green economy transition, several governments are considering the removal or phasing out of harmful fossil fuels subsidies. This policy intervention seeks to reflect the real market price of fossil fuels, thereby removing artificial barriers to the development of, among others, renewable energy and more efficient transport modes.

Since subsidy removal is a political issue in most countries, it is essential for governments to build support for the advantages of this policy. In this sense, an integrated and systemic analysis of the causes and effects of subsidy removal can help build the support needed and create a shared understanding of the advantages and potential shortcomings of this policy intervention. A systemic analysis reveals that the introduction of energy subsidies impacts energy costs in the short term—or households' and companies' energy bills—and supports GDP growth. In the medium term, a higher GDP pushes energy demand higher, increasing energy costs. This indicates that the initial gains created by the introduction of subsidies are potentially eroded over time, with the added burden of higher government expenditure (the subsidy itself). Conversely, fossil fuel subsidy removal would increase energy costs, which could possibly have negative impacts on GDP and consumption in the short term. On the other hand, medium term impacts will be positive, as higher costs reduce demand and stimulate investments (e.g. in energy efficiency). Consequently, the country, or a company, would have lower costs, thus becoming more competitive, and would become less vulnerable to energy price fluctuations.

As indicated above, fossil fuels subsidy removal is likely to bring environmental, social and economic benefits in the medium and longer term. On the other hand, several case studies demonstrate that there is no universal approach to subsidy reform, and that its design and implementation are not trivial. In particular, key entry points for action need to be identified and adapted to national circumstances, taking development priorities into consideration and anticipating possible system responses to policy intervention. In China, for example, it was possible to adjust domestic wholesale prices of gasoline and diesel to the price of a basket of crude oil products on the international market (IISD, 2013). In other countries, such as Nigeria, the full removal of fossil fuels subsidies in 2012 and the more than doubling of energy prices caused widespread protests, forcing the government to partially reintroduce the subsidies (IISD, 2013). Conversely, Indonesians generally agreed with the recent announcement of subsidy removal (IISD, 2013).

The example of fossil fuels subsidy reforms is helpful to understand the importance of a systemic and integrated analysis as a key step of policy/strategy formulation. The identification of key feedback loops allows us to fully explore the complex relations between the key variables influencing, and being influenced by, the subsidy. Furthermore, the different outcomes of subsidy reforms in various countries show that the system characterisation phase should always lead to the identification of suitable entry points for action, which must be adapted to the national context.

# 5
# Phase 3
## Strategy/Policy assessment

**Conceptual mistake #3:** The problem will be solved with the implementation of the intervention selected.

**What to do:** Identify the 'learning' capabilities of the system.

**Steps:**

1. Design potential interventions
2. Assess interventions (anticipate gaps and early warning signals)
3. Select viable intervention options and indicators

**Tools:**

- Indicators
- Causal loop diagrams
- Scenarios

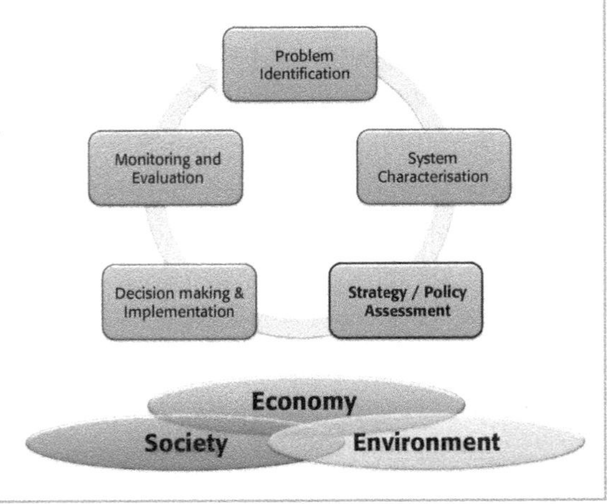

## 5.1 Conceptual mistake #3: The problem will be solved with the implementation of the intervention selected

Decision makers who adopt linear approaches to problem solving widely assume that once a valid solution has been found, the problem will be solved (Probst and Gomez 1988, 1989). For this to be true, the following basic systemic features would, among others, have to be invalidated: 1) complex systems always change over time; 2) systemic behaviour is the result of multiple, simultaneous, nonlinear interactions between various variables; and 3) complex systems' reactions to external interventions cannot be predicted (but can be anticipated).

As previously noted, 'freezing' or analysing a snapshot of a complex system is an erroneous approach to problem solving. It ignores the fact that systems evolve over time, as do problems. A solution may therefore be effective at a particular time and under certain conditions, but become totally inadequate, or even counterproductive, in a different context.

The mistaken belief that there are universal solutions to complex problems is rooted in a failure to understand the difference between complex and complicated systems. As mentioned in the introductory chapter, complicated systems comprise many different interacting parts. These parts' behaviour follows a precise logic and repeats itself in a patterned way; it is therefore predictable. When a problem arises in a complicated system, experts who know that system can find effective and lasting solutions. Conversely, dynamics that are often beyond our control dominate complex systems. These dynamics are the result of multiple interactions between variables that do not follow a regular pattern and whose dynamic interplay can lead to unexpected consequences (Miller & Page, 2007). Consequently, finding solutions to complex problems is a continuous adaptation to changing circumstances.

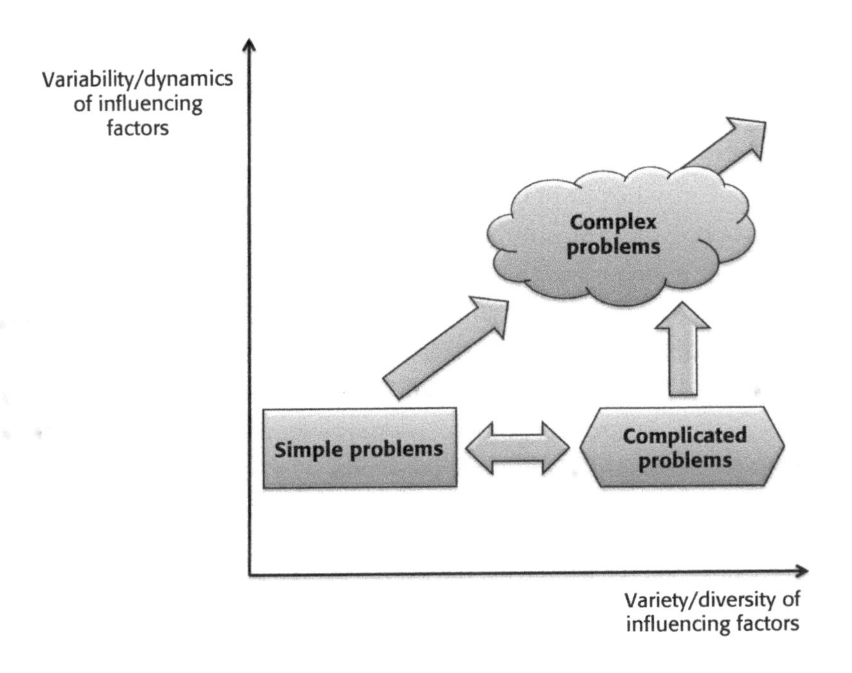


FIGURE 10 The spectrum of process complexity

Applying a solution to a problem without considering a system's potential future responses in the short, medium and long term runs the risk of aggravating the problem further, or of creating new problems that may be even more difficult (or costly) to solve. Since 'the system always kicks back' (Gall, 1977), solutions should be constantly adapted to its changing conditions. In other words, if solutions are implemented without testing the system's 'learning' capabilities, they are likely to fail.

The approach that most EU governments followed to solve the 2008 financial crisis is an example of unsustainable solutions to complex problems. The decision to prioritise debt reduction through public spending cuts and increased taxation further reduced growth, which meant only the taxpayers would pay for budgetary parity.

In the short term, the solution worked. However, in the medium and longer term, it made the situation harder to fix—also because national politics made implementing the policies for economic growth in combination with tax increases impossible in most countries.

***An example:*** the marketing of Nestlé's infant formula is another example of a decision taken without considering the system feedbacks. Nestlé's infant formula was developed in the 1920s and mainly commercialised in developed countries as an alternative to breast feeding. After the 1950s, the product experienced a sharp drop in global sales, which was primarily due to the declining birth rate in the United States. In response to this drop in sales and profitability, the corporate managers decided to introduce the product to developing countries. An aggressive marketing campaign quickly obtained excellent results. However, the comparatively high price of the product in these countries resulted in mothers diluting the formula (with possibly unsafe, non-potable water) to save on costs—which Nestlé had not considered or expected. In addition, an NGO published a report identifying the harmful effects of the diluted formula on children's health. Within this context, Nestlé decided to ignore the increasing protests. The company continued selling the product at the same price and blamed its customers for using the formula incorrectly. As a result, protests spread to the rest of the world. A growing number of consumers boycotted Nestlé products—not only the formula. This had serious consequences for the company's reputation. It was no longer considered a socially conscious firm and lost the loyalty of a part of its customers (Nutt, 2002).

## 5.2 What to do: Identify the learning capabilities of the system

The evaluation of a system's response to interventions should be based on the strategy elaborated during the integrated decision-making process. More specifically, the impacts of the selected interventions need to be measured and evaluated, paying particular attention to system-wide effects. Delays and nonlinearities have to be carefully considered at this stage in order to evaluate the system's 'learning' capabilities in the short, medium and longer term to anticipate the potential side-effects and to design a package of interventions that is more likely to create synergies and avoid side-effects.

***An example:*** the Farmer-Managed Natural Regeneration (FMNR) project in Niger is a specific example of progressively adapted interventions based on the system's learning capacity. During the 1950s and 1960s, the rapid deforestation of land for agricultural purposes resulted in severe desertification. The policy-makers decided to implement tree-planting activities to reverse the worrying trend. However, conventional tree planting to combat desertification had only limited success. This was due to two main reasons. First, tree planting in dry areas has to be carefully planned, taking the characteristics of the soil and the adaptability of tree species to the specific climatic conditions into consideration. Since these aspects were not properly addressed in the planning phase, most of the planted trees died. Second, the cost of the tree planting was too high compared with the available financial resources. The activity soon became unsustainable and was discarded. In the 1980s, a new method of reforestation, FMNR, became an increasingly popular approach to the problem of desertification. It is a successful combination of local and scientific knowledge that allows trees to regenerate from the mature roots of the previously cleared trees. This system overcomes the limitations of tree planting because it is cost-effective and uses indigenous trees that are naturally adapted to the local climatic conditions. Moreover, natural regeneration creates additional benefits for the local population as it generates new jobs and stimulates agroforestry production.

Three main steps for carrying out an integrated assessment of strategies and policies are:

1. Design potential interventions
2. Assess interventions (anticipate gaps and early warning signs)
3. Select viable intervention options and indicators

### 5.2.1 Step 1: Design potential interventions

When examining the intervention options that may influence the system and solve the problem, this investigation should be based on the entry points for action that were identified during the last step of the system characterisation phase. This allows decision makers to consider the driving forces and all the leverage points of the system. These forces include the negative and positive feedback loops already indicated in the causal diagram.

Generally, there are three main categories of possible interventions that public authorities and corporate leaders can implement in order to influence the course of events and modify a system's future behaviour. These are:

- Capital investments
- Incentive/disincentive measures
- Laws, standards or regulations

Capital investments are the most direct intervention to stimulate and accelerate change within a system. They comprise an injection of capital, which originates from the budget allocated to fund an activity that is likely to improve the system's overall performance.

> ***An example:*** at the government level, this policy intervention is preferable if the initial costs of an action are particularly high, or if the intention is to lead by example (e.g. green procurement).
>
> In the context of private companies, managers can decide to invest new resources (human or capital) in order to improve the company's performance by injecting new capital or reallocating the budget. EDF—an integrated energy company in the United Kingdom, with operations spanning electricity generation and the sale of gas and electricity to homes and businesses throughout the UK—has decided to invest in weather forecasting in order to reduce the possible negative impacts of climate change on its activities, and to maximise related opportunities. For instance, EDF uses weather forecasting to make projections regarding energy demand for cooling, based on expected average temperatures. On the other hand, the negative impacts of high temperatures are also considered, such as the possible closure of hydroelectric plants due to water scarcity (Agrawala *et al.*, 2011).

Public authorities can use incentives and disincentives—such as taxation and the granting or removal of subsidies—to influence the market and to stimulate or dissuade private investments in keeping with their policy's final objectives. Private companies can also adopt incentive measures to trigger behavioural change within the corporate environment. In particular, incentive programmes are increasingly used to motivate employees to improve productivity and enhance their knowledge and skills.

> ***Examples:*** investments in renewable energy can be stimulated by introducing feed-in tariffs, an incentive that allows households to sell the excess energy produced and increase their return on investment.
>
> Private companies provide their employees with incentives with the objective of stimulating them to remain with the company for the longer term. These incentives include, among others, the possibility to attend academic courses such as MBAs that will increase these employees' knowledge and improve their skills, ultimately fostering their career paths. Private companies may also give them the option of doing telework, thus adopting a flexible work schedule, which allows them to take care of their families and work. The employees therefore do not need to decide which of the two has to be prioritised.

Finally, public authorities are entitled to promulgate laws and establish standards to ensure that a specific objective is reached by requiring compliance from all the actors. In the private sector, company managers have the power to change internal rules and procedures, as well as to set standards of performance for their products (e.g. for a company such as Whirlpool this may mean defining an energy-efficiency target for all white appliances to be designed and commercialised in the future).

> ***An example:*** several countries have adopted renewable energy standards (RES). This standard requires utilities to generate a certain percentage of their supply from renewable energy by a specific year. Without this target, which is mandated by law, utilities would normally invest in expanding the cheapest option for power generation capacity, regardless of its carbon intensity.

Similarly, regulation in the energy sector includes fuel efficiency standards, such as those implemented in the EU and USA, to mandate yearly improvements in the efficiency of passenger vehicles and appliances. The aim is to modernise the stock and reduce energy consumption and costs, regardless of consumer demand.

***An example:*** Walmart announced a new environmental strategy in 2005. The new strategy, which gives central priority to sustainable products, implied a substantial change in internal selection and marketing procedures. In particular, the company uses a 'Sustainable Product Index', which combines indicators of energy use, material efficiency and working conditions, to assess the environmental impacts of the products it stocks. A labelling system informs the customer about the product sustainability.

In general, the implementation of new rules and the enactment of mandates ensure that the stated goals are reached and expenditure is controlled. The target is known and the required investment can be estimated relatively easily. Merely implementing incentives supports cost sharing across key actors, but does not ensure that goals are reached, which means that the costs cannot be forecasted. In the case of tax deductions for installing solar panels, for instance, the incentive (government expenditure) is entirely based on the population's buy-in. And this is only known at the end of the incentive programme.

If the strengths and weaknesses of each intervention option are considered, it becomes clear that various alternatives can be used to reach the same goal: investments, regulation and incentives in isolation or as a package, also when implemented synergistically across sectors.

***An example:*** the food industry is beginning to implement measures to control the risks of environmental changes, which can be synergistic with adaptation and may increase its resilience to climate change risks. In fact, climate change is an opportunity for companies that produce climate-resilient agricultural products and services. In particular, the commercialisation of drought-resistant seed varieties is proving successful for companies such as Monsanto, BASF, Syngenta and Bayer, at the same time improving the adaptive capacity of the agriculture sector to changing climatic conditions.

On the consumption side, companies are increasingly aware of the synergies that can be created by investing in sustainable agricultural products. For instance, Unilever, a consumer goods manufacturing company, launched a Sustainable Living Plan in 2010, according to which 100% of its agricultural raw materials should be purchased from sustainable supply chains by 2020. The Plan was effectively implemented: already by the end of 2010, sustainable agricultural raw materials amounted to 10% (Agrawala *et al.*, 2011).

### 5.2.2 Step 2: Assess interventions (anticipate gaps, time frames and early warning signals)

Prior to implementing selected intervention options, decision makers should use the available tools (e.g. scenarios and simulation models) to 'test' a system's responses to external pressures. For example, at this stage, projections can be made to create different 'what if' scenarios. These scenarios can detect possible gaps that may prevent the stated goals from being achieved. Understanding the dynamics of the system allows the identification of potential barriers to achieving the desired objectives, which should be addressed with external interventions. In this respect, if an action produces an undesirable effect on a specific indicator, it is possible to envisage—on the basis of knowledge of the structure and the system dynamics, for example the delays and non-linearity—what the future effects on other, related variables in the causal map will be. This allows decision makers to anticipate the system reactions to the implemented strategy/policy. These reactions inform decision making regarding the complementary and/or alternative actions that can be taken.

**Which steps in the decision-making cycle can scenarios inform?**

Scenarios are best used to support the first phases of the integrated decision-making process: agenda setting and the strategy/policy assessment. Scenarios are primarily oriented towards the analysis of emerging future trends, and inform activities that look forward rather than backward.

More specifically, scenarios are elaborated to explore possible future paths and trends, analyse patterns of behaviour and drivers of change in order to ultimately elaborate interventions that will allow our system to adapt (e.g. to cope with challenges and/or profit from new opportunities).

Consequently, scenarios are useful in the agenda-setting process: if an emerging trend is observed and new issues have to be faced, the company's agenda or the government's agenda will change to incorporate new concerns.

Further, since scenario analysis includes the exploration of the system responses to external or internal (man-made, such as policies) events, it can also contribute to the strategy/policy assessment. At this level, the main question is: *what are the implications of each scenario for the business/country*? This question points to the decisions to be made and possible management changes that will have to be implemented to cope with the system's expected evolution under each scenario. Despite the qualitative nature of scenarios, concrete interventions should be proposed to allow indicators to be identified and trends to be monitored in order to anticipate future challenges and opportunities.

***An example:*** policy-makers may set a target to achieve an improved food security and nutrition level within five years. Investments in the expansion of agricultural land may be identified as the best policy option to increase the annual crop yield and, thus, the availability of food. However, trend analysis and scenarios may reveal that, due to climate-change-induced rainfall fluctuations and increased water demand from the growing population, the water shortage will worsen in the same time period. Specific policies, such as investments in micro-irrigation systems, combined with the provision of water efficiency incentives, could be adopted to reduce water stress and ensure that the initial policy option is effective.

***An example:*** policy interventions to reduce $CO_2$ emissions may include incentives to cultivate sugar or starch crops—such as corn or sugarcane—to produce energy from biological carbon fixation and reduce fossil fuel exploitation. Public policy support for biofuel production has led to considerable private sector investment in the sector in several countries during the period 2006–2010. However, a systemic analysis would have highlighted emerging trends, such as the shift in land use from food crops to biofuel crops, resulting in potential increases in food prices. The analysis of these early warning signs could have indicated that policy-makers should reduce their support for biofuels, thus limiting the profitability of the sector. In this respect, companies in this market segment would have invested in technology to improve efficiency and lower the costs, or would have limited their reliance on credit. Ultimately, the loss of public support for biofuels forced several companies out of business, especially in the Unites States. This highlights the potential of using a multi-stakeholder and systemic approach to mitigate risks.

Certain system characteristics need to be closely monitored for a better understanding of how to tackle a complex problem, such as delays, effectively.

There are time delays in all kinds of systems. Often an intervention will only produce effects many years after it was implemented. Consequently, it is essential to estimate delays when deciding between different intervention options. Sometimes, the most effective actions are those that take longer to produce the desired effects; other times, however, it is necessary to choose options that guarantee an immediate effect given the urgency of the problem. Once the intervention has been implemented, the actual delay can be monitored to see if the expectations were correct, or if other feedbacks gained strength, thus changing the direction of the system.

***An example:*** in the development plans of least developed countries, policies to increase the adult literacy rate are often given lower priority than interventions that directly promote economic growth. This is primarily due to the long delay characterising the education sector and the underestimation of the contribution that knowledge can make to economic growth. Indeed, the impacts of education policies would normally only become visible well beyond the duration of a government mandate, which is also the timeline with which national targets (e.g. regarding economic growth) are set. On the other hand, improvement in education has been proved to have several positive impacts across sectors. Such investments are particularly relevant in the framework of the transition to knowledge societies. The rationale for this is presented in Figure 11. In this context, unless investments in the education sector are planned according to the timing required to observe positive impacts—in the example provided, 12 years in total for training, school time and graduation, followed by experience in the job market—more conventional capital investments will fall short of delivering the expected growth. Further, Figure 11 shows that if targets are set to transition to a knowledge society (e.g. for the growth of the services sector), but early warning signs indicate that the number of graduates in national schools is decreasing, investments should be planned that consider the delay time in the system.

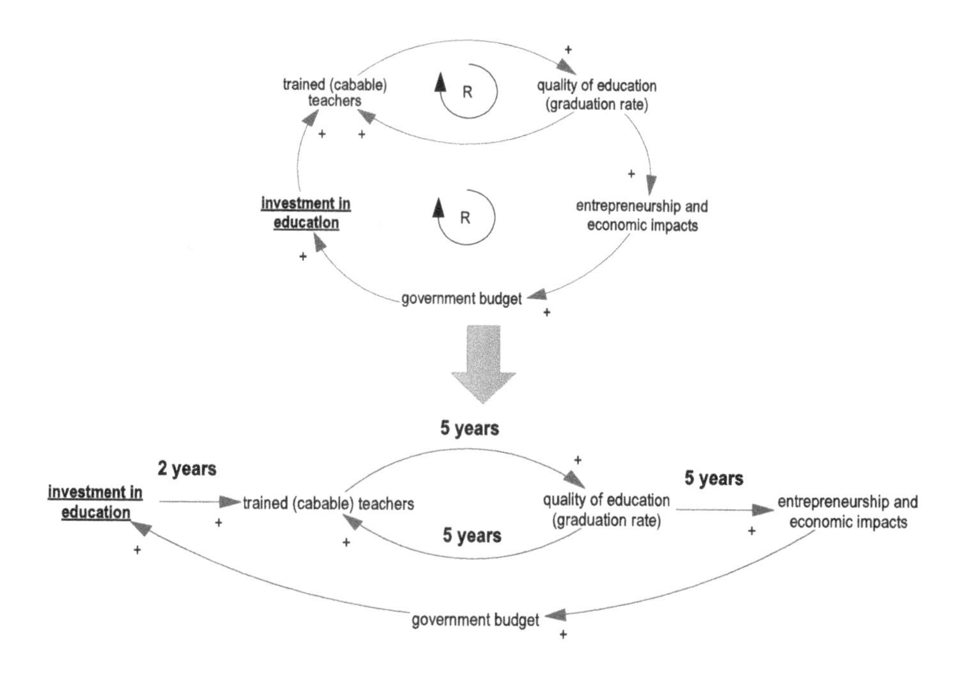


FIGURE 11 A simplified representation of the rationale for transitioning to a knowledge society, with education contributing to economic growth, also accounting for the delay time embedded in the system

> ***An example:*** the same logic can be applied to the rapid social transition that took place in South Africa, especially at the government level, after the end of apartheid. A similar case can be made for research projects at the university level, with the difficulty being experienced in finding talented researchers in STEM fields (science, technology, engineering and mathematics).

> ***An example:*** private sector investments in sustainability may imply high upfront costs and delays before significant returns are produced. However, the introduction of resource-efficient production technology and sustainable production inputs is likely to considerably improve a company's overall performance in the medium to long term. For instance, Interface, a leading producer of industrial floor coverings has reduced greenhouse gas (GHG) emissions by 50% and energy consumption by 33% in five years through a combination of targeted investments and changes in internal procedures. In addition, owing to the adoption of sustainable waste management principles (i.e. recovery and re-use), the company saved about US$300 million in ten years (Fiksel, 2006).

Another aspect that needs to be considered when evaluating the impacts of the adopted interventions is that long time delays often mean that doing and undoing may require different lengths of time.

> ***An example:*** economic and political interests often influence power generation investments. Both change rapidly. The economics of a technical solution (e.g. thermal generation vs. renewables) is highly reliant on the price of fossil fuels, among others. Political interests are based on current public concerns and national plans, but the investment is likely to be irreversible for approximately 30 years (i.e. the life time of the capital built and installed). As a result, if the price of fossil fuels changes considerably within the next 24 months, what may seem to be a good/bad economic option now, may suddenly turn into a 30-year-long economic burden/gain.

### 5.2.3 Step 3: Select viable intervention options and indicators

Systemic thinking allows us to identify and use several options to solve a problem. Among the wide array of possible options, the ones that maximise benefits across sectors and actors at minimum social, economic and environmental costs should be selected.

***An example:*** adopting energy-efficient technology requires upfront investments (capital expenditure), but will reduce energy consumption and expenditure (avoided costs), while possibly creating new jobs and income (added benefit).

***An example:*** the North Coast Forest Conservation Program and the Conservation Fund are collaborating on a project for the protection of the North Coast of California, a unique eco-region with a rich biodiversity and economically productive forests. The challenge was to identify a solution to reduce intensive deforestation in order to protect ecosystems, simultaneously maintaining the profitability of the forests. After a careful analysis of the available options and related costs and benefits, the Conservation Fund decided to purchase the Big River and Salmon Creek Forests through an innovative combination of state grants and a low-interest loan. Thanks to creative financing, the Fund reduced timber harvesting, replacing it with a diversified set of activities, including 'light-touch' timber harvests and the sale of carbon offsets, which generate more stable revenues while protecting the natural environment. The two forests are currently maintained as financially viable. Furthermore, sustainable management practices have contributed to protecting wildlife and strengthening essential ecosystem services, such as water quality protection and climate change mitigation (UNEP, 2011b).

A cost–benefit (or multi-criteria) analysis should be carried out for specific projects, and the advantages and disadvantages of public policy should be assessed. This analysis or assessment should start with the estimation of the economic, biophysical, social and cultural damage resulting from inaction. Thereafter, the costs related to implementing the different options have to be estimated (e.g. the planning, capacity building, research, operation and management). It should be kept in mind that some costs may be prolonged, or even become regular (e.g. operation and management).

In the context of green economy policies, the relevant costs that should be considered are, among others: the costs of crop losses due to extreme weather events (expressed as income and production losses, increased imports, as well as nutrition and relative health impacts) and the costs of health treatment for respiratory diseases (expressed as the number of people hospitalised, the cost of treatment and the impact on GDP through the loss of productivity).

In the private sector, the cost of inaction with regard to a product that is losing market share includes: a decline in revenues and profitability (due to lower volumes of production), a reduced consumer base (which lowers post-sale services and the potential success of the product's next update), and the potential decline of the company's reputation and branding.

**Once costs have been estimated with the help of adequate indicators, the benefits should be identified, which includes considering the direct and indirect advantages across the various system elements.**

Continuing the green economy example: the economic, social and environmental benefits should be measured. This includes, for example: measuring revenues from forest products; the avoided costs of replacing watershed management and other ecosystem services; increased revenues from eco-tourism activities due to the sustainable management of forests; increased agricultural production and value; increased food exports (or decreased imports); the avoided costs of nutrition-related diseases, or diseases related to water pollution; reduced fossil fuel costs; the avoided costs of respiratory diseases due to polluted air; increased employment; and less volatile electricity prices due to investments in renewable energy.

***An example:*** an ecosystem valuation study was conducted by Aggregate Industries UK in order to estimate the benefits of wetland restoration in North Yorkshire. More specifically, the project involved the creation of a lake and a mix of wetlands to restore the wildlife habitat. The study, using a 50-year time horizon, concluded that the benefits would far exceed the costs, as well as potential revenues from other land uses. Specifically, the total benefits were estimated at US$2 million, which comprise the value of the biodiversity, recreational activities associated with the lake, and the increased flood storage capacity. Wetland restoration would thus produce far higher benefits than the then current use of land for agriculture (TEEB, 2010).

### Box 3 Proposed tools: causal loop diagrams

A causal loop diagram (CLD) is a map of an analysed system, or, better, a way to explore and represent the interconnections between the key indicators in the sector, or in the analysed system. By highlighting the drivers and impacts of the issue to be addressed and by mapping the causal relations between the key indicators, causal diagrams shed light on possible future trajectories that a given decision generates from within, or as a reaction to external events. CLDs are more valuable if they are created in a multi-stakeholder process. They then truly help unite the ideas, knowledge and opinions of the team. This allows all the stakeholders to reach a basic-to-advanced knowledge of the systemic properties of the analysed issues.

In this context, the role of feedback loops is crucial. Feedback loops are circular causal relations among variables. The identification of negative loops (balancing the forces in the system) and positive loops (amplifying the effect of an intervention) in the causal diagram is important to identify entry points and make use of synergies emerging within and across the key elements of the system, as well as to avoid possible unintended consequences.

Causal diagrams support the decision-making process in several ways and provide valuable input during each step:

- In the **problem identification** phase, causal diagrams help identify the causal chain that determines the problem to be solved
- In the **strategy/policy formulation** phase, CLDs facilitate the identification of the key entry point for interventions
- In the **strategy/policy assessment** phase, causal diagrams support the evaluation of selected interventions: 1) short vs. long-term; and 2) direct and indirect impacts and responses
- In the **decision-making/implementation** phase, CLDs increase stakeholder participation in the definition of the integrated strategies and action plans
- In the **monitoring and evaluation phase**, the behaviour of the system can be analysed with the help of causal diagrams as they identify the synergies and/or unintended consequences of implemented interventions

*Continued*

| Strengths | Weaknesses |
| --- | --- |
| • Facilitate a multi-stakeholder approach to problem solving<br>• Help highlight causal relations between indicators<br>• Support the analysis of system behaviour and its reaction to external interventions<br>• Enable quantitative monitoring and the evaluation of implemented actions | • Effectiveness is directly linked to the quality of the process<br>• Wrong or partial causal diagrams may lead to ineffective (or even harmful) interventions<br>• Best used if combined with quantitative tools (e.g. simulation models) |

**TABLE 4 Strengths and weaknesses of causal diagrams**

Decision makers can combine causal diagrams with other tools proposed in this book to maximise the benefits throughout the problem-solving process. More specifically, the identification of key variables and entry points for intervention allows indicators to be directly used and analysed if a systemic approach is used when creating CLDs. Causal diagrams also complement influence tables by adding 'dynamics', so that, beside the interconnections across the key variables, the patterns of behaviour can also be analysed. Finally, these diagrams support the creation and analysis of scenarios, and are necessary for the correct development of integrated and dynamic simulation models.

## Case study 3 Reducing risks and maximising profits with systemic strategy assessment: Nestlé

From 1997, under the new leadership of Peter Brabeck, Nestlé adopted an integrated strategy for the management of its global business (Probst *et al.*, 2008). On the one hand, managers analysed the current operational successes achieved throughout their global presence and derived lessons learned from different market structures and geographical contexts. On the other hand, they built scenarios of possible future business developments, in order to assess their current strategies in light of potential upcoming challenges and opportunities.

Based on an analysis of the future trade scenarios, Peter Brabeck warned that Nestlé generated about 70% of its sales in mature markets with very low growth potential. New markets and sources of profit had to be found in order to achieve the corporate target of a stable and organic 4% annual growth rate. Once, based on an understanding of the current and future market dynamics, the problem had been identified, a number of interventions were proposed to address the issue and achieve the expected results. More specifically, the company decided to invest in a variety of programmes to strengthen its long-term profitability, including product innovation, the reduction of packaging costs, the adoption of new efficiency standards, investments in technological improvements, expansion into new market segments (such as wellness products) and the standardisation of electronic data, among others (Probst *et al.*, 2008).

This comprehensive approach allowed Nestlé to obtain considerable returns. Despite the food industry's average annual market growth of 2%, Nestlé increased its sales in 1997 from CHF70 billion to CHF98 billion, and its profits from CHF4 billion to CHF9 billion.[5] In 1997, the Group's organic growth averaged 5.8%, which was significantly higher than that of its main competitors. In addition, by consulting more than 400 decision makers from over 40 countries, Nestlé's national and regional contexts adopted

5 CHF1 = US$1.11, exchange rate as of October 2013.

its global efficiency standards. In each of the three main regions (North and South America, Europe and Asia) which the company targeted, pilot projects had previously tested and assessed their standards. The success of these initiatives speaks for itself: total savings of more than CHF12 billion were achieved. Consequently, between 1997 and 2006, Nestlé's margin increased from 5.7% to 9.3%, and its cash flow doubled (Probst *et al.*, 2008).

As this case study demonstrates, an integrated and systemic approach to strategy/policy assessment should be based on the analysis of current and future challenges and opportunities, and supported by relevant tools, such as scenarios. In other words, the assessment should consider the short, medium and long-term impacts of the strategy and be based on the previous exploration of the system's dynamic properties. When Peter Brabeck decided to invest in innovation and efficiency, Nestlé was the world's leading food company, and there was no apparent reason for a strategic change in the short term. However, the adoption of systemic thinking identified early warning signs and possible threats for the company's long-term profitability. Also, the success behind Nestlé's strategy relied on the combination of different simultaneous and complementary interventions. The multiple programmes implemented were based on an understanding of the interactions between the different corporate priorities and were developed to maximise synergies and avoid unintended consequences.

# 6

# Phase 4

## Decision making and implementation

**Conceptual mistake #4:** With a problem-oriented optimisation, the solution will maximise benefits for all.

**What to do:** Evaluate the proposed solution using different perspectives, and assess their impacts across sectors and actors.

**Steps:**

1. Use a multi-stakeholder approach to assess roles and responsibilities
2. Analyse the expected impacts across sectors and actors
3. Define the strategy/policy and action plan

**Tools:**

- Indicators
- Causal loop diagrams
- Scenarios

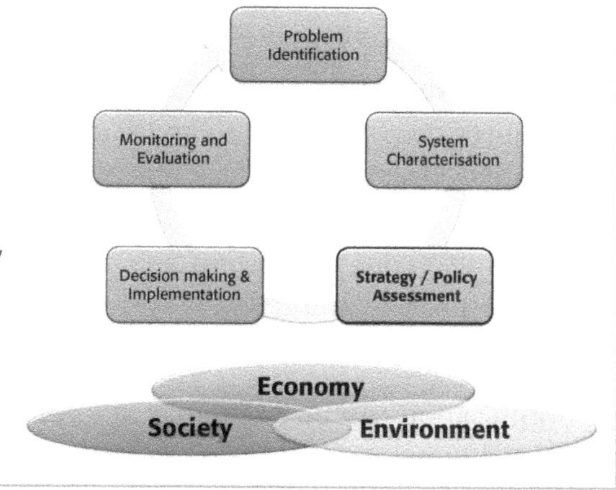

## 6.1 Conceptual mistake #4: With a problem-oriented optimisation, the solution will maximise benefits for all

Decision making is a challenging task that, if not properly managed, can lead to the emergence of unexpected side-effects and failure. Common mistakes in the decision-making and implementation phase are: focusing exclusively on the problem to be solved; implementing actions aimed at maximising the performance of a few indicators or sectors; and assuming that all the actors will benefit from the solution. By doing so, decision makers pay little or no attention to the complex relations between the different elements of the system. The assumption that, once implemented, the solution will benefit the entire system, which will thrive again, is erroneous.

Conversely, an integrated approach to problem solving recognises that optimisation is always limited, especially if the role and goals of the different actors involved in complex problems are taken into consideration. Implementing strategies designed to maximise the performance of part of the system, or to solve the problem of one (or a small group of) actor(s), generally leads to resistance from other stakeholders.

> ***An example:*** the Government of Mauritius recently considered the construction of a coal-fired plant to generate electricity. The Central Electricity Board suggested this after carrying out an analysis focused on minimising the electricity supply costs. Strong public opposition due to concerns about $CO_2$ emissions halted all plans for the expansion of the power capacity—especially in terms of thermal generation—in March 2013.

Optimisation can be compared to a compass. It tells travellers where north is, but makes no reference to the obstacles or aids, such as mountains and lakes, that they will encounter on the way. Similarly, optimisation leads decision makers to design problem-oriented strategies. These strategies strive for immediate results (i.e. heading north) but ignore the key variables, such as systemic structures, feedbacks, delays and nonlinearities (i.e. the obstacles or aids along the way), which play a crucial role in the strategy's success.

## 6.2 What to do: Evaluate the proposed solution using different perspectives, and assess the impacts across sectors and actors

When selecting the most effective intervention options, various points of view and perspectives should be considered. Their likely impacts across sectors and actors should also be assessed. In this context, a multi-stakeholder approach becomes crucial.

On the one hand, the main objective of the strategy/policy is to solve the complex problem that drew the attention of the decision makers and triggered the decision-making process. On the other hand, the options that improve the system's overall performance should be prioritised above those that only address a specific problem. This is crucial to identify and implement lasting solutions that effectively solve complex problems and avoid side-effects.

Returning to the above analogy: a systemic approach to decision making provides decision makers with more information than a compass can. By relying extensively on a multi-stakeholder approach, by being able to anticipate gaps, and by its emphasis on risk mitigation and resilience, the systemic approach functions like a GPS. It provides a series of itineraries from which to choose. It also allows the traveller to consider the total distance, timing, toll roads, etc. Further, while a compass can only indicate one direction, the GPS informs decision making by identifying various possible paths (or itineraries), each of which is based on specific priorities and goals, and also indicates each option's potential drawbacks and risks. This approach is complemented by additional support features, such as identifying potential short vs. long-term challenges that may arise from the complexity of the system (e.g. system responses based on its learning capabilities).

***An example:*** the Madrasati initiative, launched by Queen Rania Al Abdullah in Jordan, is a virtuous example of a multi-stakeholder decision-making process. The initiative seeks to create shared responsibility among all the relevant stakeholders involved in the education sector with regard to the restoration and maintenance of schools' infrastructure and the improvement of the learning environment. In the case of the Husban Secondary School for Boys, for instance, Madrasati collaborated with the Central Trade and Auto Co.–Toyota, which acted as a private sponsor, as well as with other major stakeholders. They ensure that the teachers, administrators and students properly maintain the repaired infrastructure and new equipment (e.g. printers). A wide variety of stakeholders were therefore involved in the elaboration of a strategy for Husban School. This strategy focuses on the main needs and the relevant stakeholders are given different responsibilities. It was essential to ensure that the participants understood that the donor (i.e. Toyota) was not a 'fairy godfather' who provides for all needs without asking anything in return. Rather, Toyota is a partner that offers genuine cooperation with all the parties responsible for the school's future. The experiment proved successful. It has not only succeeded in having different perspectives included in decision making, but has also fostered the community's ownership of the school (Stadtler *et al.*, 2010).

An example of the importance of multiple perspectives and ideas for business development is the success obtained by Procter & Gamble (P&G) with its open innovation programme, 'Connect + Develop'. This programme consists of an Internet platform where all companies and entrepreneurs can propose new innovative business solutions to meet the company's needs. This open approach to innovation provides the company with a variety of different options, from which the most valuable are chosen and developed in collaboration with the proposing partner (Zimmermann *et al.*, 2013).

Effective decision making and implementation require the following main steps:

1. Use a multi-stakeholder approach to assess roles and responsibilities
2. Analyse the expected impacts across sectors and actors
3. Define the strategy/policy and action plan

### 6.2.1 Step 1: Use a multi-stakeholder approach

After the problem has been identified as part of a broader system and different intervention options have been evaluated, a rigorous logic should be followed to define the integrated strategies/policies. This starts by agreeing on the general principles that apply to the analysed system. It continues with defining the strategic/policy guidelines for intervention and ends when the action plans have been determined. The plans are built up of complementary, cross-sectoral intervention packages targeting different elements of the system in which the problem occurs.

All of the relevant stakeholders should first be included in the decision-making process, and agree on the general principles and goals, before an integrated strategy/policy (e.g. key indicators) is developed. Furthermore, the stakeholders should agree on the tools they wish to use to formulate and evaluate the strategy. Their discussions should be based on their common understanding of the system structure and dynamics (e.g. by means of the causal loop diagram), as well as on the basic conditions required for change to happen. At this stage, it is crucial for the stakeholders to reinforce their common understanding of the need for concerted actions. If discrete interventions are not aligned with the overarching strategic/policy guidelines and goals, they are likely to disregard the systemic complexity, which would mean ignoring the cross-sectoral causal relations. This would lead to unintended consequences and, possibly, failure.

> ***An example:*** ecosystem degradation could be a worrying trend in a country whose economy relies heavily on tourism activities. After having explored the dynamics of the system with the help of the relevant tools—indicators, influence tables, causal maps, scenarios and simulations—policy-makers and other stakeholders should agree on the general principles for intervention. They should, for example, acknowledge that the ecosystems and ecosystem services are an essential economic, environmental, social and cultural public good and component of the national identity. Thereafter, they should identify specific strategies/policies to address the problem of ecosystem degradation and its relation to the tourism industry's poor performance.
>
> The following are examples of strategies:
>
> 1. Provide national guidance regarding protecting existing critical ecosystems, existing coastal development, and future investment
> 2. Engage the tourism sector in climate change adaptation activities for more sustainable development
>
> Each strategy can then be broken down into concrete actions with each stakeholder being assigned clear responsibilities.

The first strategy could lead to the following actions:

(a) The relevant authorities develop a guideline to be used at the national and local levels to prepare climate change adaptation plans for the tourism sector

(b) The relevant authorities, with the support of the national and international tourism organisations and entities, develop three pilot studies to examine the efficacy and utility of the guidelines mentioned above

The second strategy can be implemented by means of the following actions:

(a) The tourism authorities, in collaboration with other relevant bodies, promote micro-finance schemes to support tourism adaptation and sustainable development programmes

(b) Coastal and wetland attractions, as well as non-coastal attractions, are developed to demonstrate how climate change adaptation mechanisms can be used

(c) In coordination with ongoing activities in the fisheries sector, the tourism sector can help with the economic assessment of the coastal and marine ecosystems, in addition to its existing efforts to involve the community in development projects in order to preserve the ecosystem

While clear roles and responsibilities have to be allocated in the implementation phase, it is essential that all the relevant actors collaborate in all the steps of the decision-making process. Designing and implementing a comprehensive, cross-sectoral strategy/policy require different perspectives, specialised knowledge and assorted skills to make use of the synergies and avoid side-effects.

More specifically, a multi-stakeholder process (MSP) can help ensure better coordination between the actions (and their impacts), which may overlap or be contradictory. In addition, an MSP ensures that all the knowledge is merged and properly utilised by means of common mental models. In particular, integrated decision making should involve the participation of actors from:

- Different levels (e.g. international, national, regional and local; or different management and responsibility levels in a company)
- Different sectors (e.g. economic, social and environmental; or different branches in a company)
- Different interests (e.g. public, private, civil society)

***An example:*** defining an integrated strategy to address the problem of unbalanced energy supply and demand in a country requires the participation of different kinds of actors. The stakeholders range from those with technical to political profiles to those with expertise in different fields. They are also a balanced representation of the public and private sectors. In particular, a team of technical experts can recommend energy-efficiency technology options, while the strengths and weaknesses of these options in different sectors may be analysed with the help of sectoral experts, i.e. specialists in manufacturing processes, transport, urbanism, etc. Economists can advise on the estimation of the costs and benefits, as well as on possible incentives, i.e. tax reductions, subsidies, etc. Finally, the policy-makers have to design an integrated strategy to bring the different perspectives together and to take the potential delays, feedback loops and nonlinearities of the different intervention options into consideration. As a specific application, Chevron is collaborating with the public sector to find effective solutions to energy sustainability issues. To do so, the company decided to establish Chevron Energy Solutions (CES), a provider of energy-efficient facility upgrades, directly funded by the energy savings (i.e. avoided costs) accrued by its clients. CES public-sector projects aim at reducing resource consumption, while also avoiding GHG emissions and energy costs. In 2005, CES projects saved 34.0 million cubic metres of natural gas and 177 million kWh of electricity use, and avoided 97,000 metric tons of carbon dioxide emissions (Fiksel, 2006).

A practical example of an approach that uses stakeholder consultations in order to solve problems is that of the UNDP regarding Guyana's low carbon development strategy (LCDS). A multi-stakeholder steering committee guided the formulation of the LCDS. This committee comprised members from different levels, sectors and interests. They included (UNDP, 2012):

- International organisations (e.g. WWF, International Institute for Environment and Development)
- Relevant national ministries (e.g. Ministry of Agriculture, Ministry of Amerindian Affairs)
- Private sector associations (e.g. the Forest Producers Association)
- Indigenous groups associations (e.g. The Amerindian Action Movement of Guyana)
- NGOs (e.g. the Federation of Independent Trade Unions of Guyana)
- Representatives of women's and youth's rights and interests

The ideal MSP follows a five-step approach (Hemmati *et al.*, 2002):

1. Setting the context for collaboration (i.e. problem and stakeholder identification)
2. Framing the goals and objectives of the discussion (i.e. agenda-setting, group composition)
3. Providing relevant inputs (i.e. policy formulation, capacity building)
4. Organising dialogues and meetings (i.e. facilitation, decision-making)
5. Implementing outputs (i.e. action plan implementation, monitoring and evaluation)

Each step involves specific actions to ensure the stakeholders accept maximum ownership of the process and to reassure them that all actions are first discussed before being considered for incorporation into the integrated strategy. Moreover, contacts should be maintained with the non-participating stakeholders throughout the process.

### 6.2.2 Step 2: Analyse the expected impacts across sectors and actors

The expected impacts of the strategy and related interventions should be simultaneously evaluated across a variety of indicators (social, economic and environmental). The strategy's success over the medium to longer term is largely determined by the way social, economic and environmental factors are supported with solid and coherent information, and the way they interact with each other in the causal loop diagram and simulation model. It is crucial to analyse the strategy/policy impacts to better understand the drivers of change and system responses, and to design effective interventions that lead to desired impacts.

**Why use simulation?**

In the context of this book, simulation exercises are dynamic and integrated 'what if' assessments that can create insights into strategy/policy implementation's impacts.

- **Dynamic.** Indicating that feedback loops, delays and nonlinearity–the three core elements that underlie the characteristics of real systems–are explicitly taken into account and represented in the model
- **Integrated.** Indicating that the simulation model has to include cross-sectoral indicators and incorporate the economic and biophysical dimensions of the problem and analysed system
- **What if.** Indicating that models should not tell us what to do, nor how to optimise our system. Instead, they should provide an evaluation of the impacts of the decisions being considered for implementation and should test whether these can effectively solve the problem

As opposed to mainstream modelling approaches, which are neither dynamic nor integrated, the simulations presented in this book provide the following advantages:

- **Estimation of strategy/policy impacts across sectors.** The simulation of scenarios with quantitative models allows decision makers to evaluate the impact of selected interventions within and across sectors
- **Debunking complexity.** The simulation of causal descriptive models helps simplify and gain understanding of the complexity and evaluate the short vs. longer-term advantages and disadvantages of the analysed interventions
- **Prompt feedback.** Currently, the simulation of a scenario with causally descriptive models takes no longer than a couple of seconds for models with more than 15,000 equations. Consequently, an average-sized model will simulate a scenario from 1980 to 2050 in less than a second and provide decision makers with immediate feedback
- **Realistic insights.** A causal descriptive model can capture new patterns of behaviour and help identify potential side-effects and additional synergies

First of all, the impacts on the specific problem(s) that the strategy addresses need to be directly evaluated. Expected changes in the indicators representing the problem should be measured to test the actual effect of planned actions.

***An example:*** if deforestation has been identified as a worrying trend and establishing new protected areas has been selected as a policy option to protect forests and reverse the trend, deforestation and its causes are the first indicators that should be monitored. Further, if a reduced deforestation target has been set, a comparison should be made to evaluate whether the trend towards this target is sufficient to achieve it on time.

Second, the expected impacts should be analysed with the help of the causal map. It is important to remember that different key causes, which determine the success (or failure) of the interventions, influence different sectors. In other words, an action can have very positive impacts on certain sectors and create issues for others. Furthermore, in the longer term, successful interventions may have negative short-term impacts. Mitigating actions should therefore be designed and implemented.

***An example:*** policy-makers in some sub-Saharan African countries are investing in modern, climate-resistant infrastructure to increase the resilience of the transport sector. Such investments are likely to produce long-term benefits across a number of sectors:

- Improving the mobility of people and goods will boost national and international trade, which will lead to more investments, increased income and overall economic growth.
- In addition, the prices of food and industrial commodities will decline as a result of the improved transport time and costs, which will have positive effects on food security and the manufacturing sector
- Moreover, transport modes others than private vehicles (e.g. trains) will reduce the energy consumption, in particular that of fossil fuel, which will in turn lead to lower $CO_2$ emissions
- Finally, establishing green public transport networks will contribute to the greening of cities and to adaptation to the negative impacts of climate change

However, although the cross-sectoral benefits are evident in the medium to long term, high upfront costs and potential implementation delays may undermine the feasibility of the intervention in the short term.

***An example:*** the volcanic ash that spread from Iceland across Northern Europe on 14 April 2010 illustrates different short-term vs. long-term impacts. On the evidence of jet engine failure after the 1991 Pinatubo volcanic eruption, the International Civil Aviation Organisation (ICAO) decided to stop thousands of flights for a few days. On the one hand, the decision was very effective in preventing potential disasters and was supported by various stakeholders. On the other, however, the very precautionary approach that the ICAO adopted was perceived to be overly strict by other stakeholders (World Economic Forum, 2012). While extreme precaution was necessary in such a delicate situation, and proved to be very effective in avoiding fatal accidents, which is very positive for passengers, it is important to consider that this decision also had negative economic impacts on other actors in the system—the airlines—which protested strongly against the long delays before regular flight schedules could be resumed.

Moreover, the direct and indirect impacts that selected interventions may have across different actors (e.g. private companies, vulnerable groups, rural vs. urban communities) should be presented with the help of causal maps and simulations.

***An example:*** the German climate change programme for the building sector is an example of a successful policy measure that has produced positive impacts across a variety of actors. The initiative aims at providing incentives for energy-efficiency refurbishments of existing housing. Funded by the German promotional bank KfW Bankengruppe, it includes different promotional initiatives such as loans and grants targeting homeowners, private builders, landlords and housing companies. The level of financial support is proportional to the investor's energy-efficiency target (e.g. the best standard, KfW Efficiency House 55, receives the highest loan). This incentive package enables $CO_2$ emission reductions of 5 million tons every year, with a positive impact on climate change mitigation, pollution reduction and health. Moreover, energy efficiency in buildings has a direct impact on households through the reduction of energy use and costs. Finally, the programme has contributed to the expansion of local small and medium construction enterprises, thereby positively impacting national economic growth and employment (UNEP, 2011b).

***An example:*** the promotion of organic agriculture in the Republic of Moldova brought considerable benefits to a variety of actors. The success of the strategy was due to the effective combination of multiple policy measures, including regulations, subsidies and institutional development. Subsidy programmes addressing farmers and private investors complemented the adoption of an organic agriculture marketing law in 2005. Moreover, local farmers received training sessions on organic farming practices. The institutional framework was also reinforced through the establishment, within the Ministry of Agriculture, of a Department for Organic Farming and Renewable Resources. As a result, in the period 2005–2009, the organic farming area in Moldova increased from 715 hectares in 2005 to 31,102 hectares in 2009, a 45-fold increase. Together with environmental benefits from reduced use of chemical fertilisers and pesticides, organic agriculture is now an important sector for income and employment generation in Moldova, given that it is more labour intensive than conventional agriculture, and products are sold at higher prices on the domestic and international markets (UNEP, 2011b).

### 6.2.3 Step 3: Define the strategy and action plan

Figure 12 illustrates the decision-making context in which problems emerge and in which solutions need to be found in the system characterisation stage by means of a systemic approach to decision making. This approach considers: 1) the strategy/policy context; 2) the projected scenarios; and 3) the complex structure of the analysed system.

More specifically, in order to design and evaluate effective interventions, the following tasks should be carried out:

- First, the structure of the system—an integrated set of social, economic and environmental drivers—should be properly analysed and understood (see the bottom of Figure 12). The key causes of the problem, as well as its main impacts, have to be evaluated from a systemic perspective to identify the key entry points for action and the possible synergies.

***An example:*** climate change is a global problem that is observed through the monitoring of environmental indicators, such as temperature, precipitation patterns and trends. In order to design and implement effective policies to address both the causes and impacts of this worrying phenomenon, the structure of the system—which includes economic, social and environmental variables—needs to be carefully analysed. First, the main cause of climate change—the excessive amount of greenhouse gas (GHG) in the atmosphere—should be analysed to identify possible entry points for emission reduction interventions to mitigate climate change (i.e. changing production processes and technologies, or shifting to low-emission transportation). Second, the direct and indirect climate change impacts on the environment (e.g. ecosystem degradation, biodiversity loss, etc.), the economy (e.g. reduced agricultural production, economic losses due to extreme weather events, etc.) and society (e.g. the population's vulnerability due to shifting weather patterns, which includes reduced access to key services and resources) have to be fully understood in order to plan for adaptation and, ultimately, improve climate resilience and well-being.

***An example:*** Drive Now, the innovative car-sharing service developed by BMW in collaboration with a number of partners, is a valuable example of synergistic value creation resulting from a systemic approach to business development. The service was designed to address some of the key public mobility challenges, such as congestion and environmental pollution. By leveraging its internal capabilities and overseeing multiple external partnerships, BMW was able to analyse the structural causes of the problem, and to identify key market entry points. Thanks to the integration of economic, social and environmental aspects of business development, BMW was able to create public value by introducing a new sustainable solution to individual mobility, while also developing a profitable business (Zimmermann *et al.*, 2013).

- Second, possible future scenarios affecting the system (such as economic volatility, climate impacts, natural disasters and other unexpected events) have to be analysed. They can have a considerable impact on the analysed interventions' effectiveness.

Besides this impact analysis, scenario analysis and quantitative projections can also reduce uncertainty.

In keeping with the previous example: climate change mitigation and adaptation policies should be based on the analysis of the problem's possible future developments (e.g. which sectors and communities will be most impacted in the coming years?) and the projected market trends. When, for instance, assessing the suitability of energy-efficiency policies, attention should be paid to energy price scenarios. An energy-efficiency policy is more likely to be successful if energy prices were to rise. With a higher expected expenditure on energy, the private sector and households would be more inclined to invest in energy efficiency and make use of the government-provided incentives. On the other hand, if energy prices were to decline, and given the transition costs, consumers may deem a shift to energy efficient technology inconvenient and expensive. Decision makers may therefore prefer to target the energy supply and to invest in renewable energy development. Consumption behaviours can then be targeted at a more suitable moment.

- Third, the choice of intervention options should be based on the structure of the analysed system and on a variety of possible future scenarios. In order to successfully evaluate the effectiveness of the selected intervention options—and to determine whether they create synergies, bottlenecks or side-effects across sectors—a balance should be found between the main strategy/policy options. This will prioritise the system's improved performance (rather than that of its parts) and ensure that the costs and benefits are fairly allocated across the key actors.

***An example:*** Meinert Enterprises adopted a systemic approach to strategy development. Meinert is a Saskatchewan agriculture operation concerned with dryland farming, which produces cereals, pulses and forages under challenging climate conditions, due to both annual and seasonal variability and unpredictable frosts. The managers analysed the complex set of uncertain factors that play a central role in farm decisions, including climate variability and related economic risks (e.g. interest rates, energy costs, cash flow minimum requirements). In response to these challenges, Meinert Enterprises employs adaptive agricultural practices, such as crop diversification, enhanced early moisture infiltration and crop rotation to improve soil quality. Meinert also employs different strategies to reduce the financial risks related to climate change, including participation in stabilisation programmes, purchasing of crop insurance and earning off-farm income (Nitkin *et al.*, 2009).

By implementing this approach within the decision-making cycle, systems thinking facilitates the understanding of complex system structures and helps maximise strategy/policy outcomes, while avoiding unintended consequences. If we were to identify a few general guidelines for effective decision making in the context of complex problems, these would be:

- **Use a multi-stakeholder approach** to consider all systemically relevant points of view and incorporate as much cross-sectoral knowledge as possible into the analysis
- **Evaluate the impacts across sectors and find a balanced strategy**, keep the long term in mind, prioritise resilience and aim for improved performance of the whole system rather than maximising selected indicators at the expense of others
- **Evaluate the impacts across all the economic actors and find an inclusive strategy**, thus ensuring that the costs are coherently allocated and the benefits are equitably distributed across the key actors in the system.

The relation between energy policies and development objectives can be used to demonstrate the need for integrated strategies. For instance, policy-makers in a developing country who have to decide on energy strategies may favour exploiting fossil fuels, if available, to increase the energy supply at relatively low costs. However, a multi-stakeholder, three-layered approach to integrated decision making should be followed to ensure the inclusiveness, effectiveness and sustainability of the selected policies (see Figure 12).

| **Policies/ strategies** | **Investment**<br>(e.g. capital investment in renewable energy and energy efficiency for extra capacity and retrofits) | **Mandates and targets**<br>(e.g. renewable energy and energy efficiency standards, deforestation and reforestation targets) | **Incentives**<br>(e.g. feed in tariffs for energy, tax rebates, payments for ecosystem services) |
|---|---|---|---|
| **Scenarios** | Climate change, energy prices, conflicts, peak oil, world economic growth, etc. | | |
| **Structure** |  Economy / Society / Environment | | |
| | **Social sectors** | **Economic sectors** | **Environmental sectors** |
| | Population<br>Education<br>Infrastructure *(e.g. transport)*<br>Employment<br>Income distribution | Production (GDP)<br>Technology<br>Households accounts<br>Government accounts<br>Investment *(public and private)*<br>Balance and financing<br>Government debt<br>Balance of payment<br>International trade | Land allocation and use<br>Water demand and supply<br>Energy demand and supply<br>*(by sector and energy source)*<br>GHG and other emissions<br>*(sources and sinks)*<br>Footprint |

FIGURE 12 The three main layers for an integrated strategic decision-making process: structure, scenarios and policies

- First, the environmental, social and economic reality of the country should be carefully analysed. This analysis should include its available natural resources, the use of energy in key economic sectors, rural communities' access to energy, etc.
- Second, alternative scenarios should be created for different energy policies. For example, the intensive use of fossil fuels will increase $CO_2$ emission levels, thus producing mid to long-term negative impacts on the well-being of the population (e.g. respiratory diseases due to air pollution) and on the environment (e.g. due to climate change impacts on ecosystems). Further, the consumption of fossil fuels may lead to the depletion of natural reserves and to an increase in imports, making the country dependent on foreign trade and more vulnerable to price volatility.
- Third, policies should be designed to offset possible negative developments. Interventions may lower the costs and ensure a higher degree of energy self-sufficiency. If households and industries were to adopt energy-efficiency standards, this would reduce energy consumption and cut costs in the longer term. In addition, incentives to exploit renewable energy sources will contribute to a reduction in $CO_2$ emissions and will simultaneously increase domestic energy production and possibly improve access to electricity in rural areas, while also reducing fossil fuels imports. Moreover, in developing countries, which still widely use wood-burning stoves for cooking, greater access to modern forms of energy (electricity and, therefore, electric stoves) has the potential to reduce mortality due to inhalation of smoke, and to reduce deforestation.

Although implementing these policies may lead to higher short-term costs, they will maximise the benefits across sectors, thus increasing employment and energy access, while reducing natural resource consumption and emissions, and increasing resilience.

## Box 4 Proposed tools: scenarios

Scenarios are expectations of possible future events. They are used to analyse potential responses to new and upcoming developments. Scenario analysis is therefore a speculative exercise during which several future development alternatives are designed, agreed upon, explained and analysed in order to discuss what is needed to lead to them and the consequences they could have on our system (e.g. a country or a business).

Scenario analysis is designed to improve decision making. This is done by including uncertainty and risk, and analysing the patterns and paths, as well as the strategic responses. Scenario analysis can also be used to explore possible unexpected events, which will increase the general readiness to deal with unforeseen external impacts on the system.

The use of scenarios can support decision makers in different stages of the strategy/policy-making process, in particular:

- In the **problem identification** phase, emerging trends can be made visible by creating future scenarios. Consequently, the company or government agenda can be changed to incorporate new concerns
- Since scenario analysis includes exploring the system's responses to external or internal events, it can also contribute to the **strategy/policy formulation and assessment** stages
- In the **monitoring and evaluation** phase, scenarios can be used to identify interventions, if any, that may strengthen the current strategy/policy. In this sense, they help anticipate the opportunities that the existing interventions offer, which helps identify synergies and thus maximises their effectiveness

| **Strengths** | **Weaknesses** |
|---|---|
| • Promote innovative thinking about the future system's behaviour<br>• Improve multi-stakeholder, cross-sectoral risk management<br>• Support the monitoring and identification of potential future challenges resulting from implemented actions | • Inadequate for short-term planning<br>• Require continuous revision and control by a specialised team<br>• Require a disciplined monitoring, analysis and communication process<br>• Ineffective if not integrated into the decision-making process |

TABLE 5 Strengths and weaknesses of scenarios

Scenario planning is useful—even when used in isolation—to explore emerging patterns, regardless of whether external events, or developments internal to the system, trigger them. Developing scenarios requires several tools, discussed in this book, which contribute to the scenarios in various ways.

*Continued*

Causal loop diagrams (see Box 3) help formalise the mapping of the system by means of scenario analysis. They do so by clearly identifying the causal relations and feedbacks responsible for internally generated changes, as well system responses to external changes. Along the same line, simulation (see Box 5) could make use of the complementarity of causal diagrams and scenarios, turning the qualitative analysis into a quantitative assessment. Finally, scenarios make use of and contribute to the use of indicators. Scenarios are built on observations of reality by means of selected indicators. Further, by looking forward in time, new scenarios may indicate that a new set of indicators should be monitored to anticipate upcoming challenges and opportunities.

## Case study 4 Multi-stakeholder decision making in relief operations: the Agility CSR programme

An illustrative case study of a successful approach to decision making and implementation is that of Agility, a Kuwait-based global logistics company that developed a new corporate social responsibility (CSR) programme centred on the collaboration between a variety of public and private actors, and aimed at supporting disaster relief interventions in several countries (Stadtler and van Wassenhove, 2011).

Frank Clary, Senior CSR Manager at Agility, realised that a more coordinated and systematic approach to corporate disaster response was needed after analysing the difficulties encountered during the relief operations in the wake of the 2006 Lebanon crisis. In that situation, although Agility had successfully provided logistic support to humanitarian operations, Frank noticed that there was still room for improvement, especially with respect to rapid deployment procedures and communication among all the relevant stakeholders (Stadtler and van Wassenhove, 2011).

As a result, Frank implemented a strategy based on three pillars: 1) integrating the disaster relief activities into the broader CSR programme and developing internal support structures and capacities; 2) setting up collaboration with humanitarian organisations through the establishment of partnerships that last beyond a single disaster; and 3) adopting a coordinated approach with the United Nations (UN) and other leading logistics companies to align activities on the ground. This multi-stakeholder approach led to the implementation of several successful projects, such as the LETs initiative, the first multi-company commitment—involving the leading logistics companies Agility, UPS, TNT and Maersk—to support humanitarian relief efforts during natural disasters, under the coordination of the Global Logistics Cluster, led by the World Food Programme. Another successful initiative was the HELP programme, whose objective was to gather and institutionalise lessons learnt from the local employees, and to create online resources in e-rooms and blogs (Stadtler and van Wassenhove, 2011).

Overall, this innovative CSR strategy proved successful in supporting humanitarian operations in a variety of disaster situations, including floods in Pakistan, earthquakes in Chile, Haiti and Indonesia, cyclones in Myanmar and Bangladesh, among others. In each of these contexts, Agility was able to adapt to the local context and provide logistic support as needed, in close collaboration with public and private partners on the ground (Stadtler and van Wassenhove, 2011).

This case study is particularly relevant for the decision making and implementation phase of the strategy/policy cycle, since it follows the key steps for an integrated approach to problem solving. First, a multi-stakeholder process is adopted for the deployment of humanitarian assistance, requiring a close collaboration between Agility's employees, international organisations and community-based associations. This approach facilitates the comparison of different perspectives and identification of potential challenges for implementation, assigning central importance to the local context. Second, the establishment of stable and trustful partnerships with humanitarian organisations and other private companies allows the definition of shared operational procedures. This aspect is key to ensuring rapid decision making and implementation processes, which are based on the evaluation of the expected impact of the assistance deployed (*ex ante*), and a clear allocation of roles and responsibilities among the actors involved.

# 7

# Phase 5

## Monitoring and evaluation

**Conceptual mistake #5:** Monitoring and evaluation do not affect the decision-making cycle, they only evaluate the system performance.

**What to do:** Evaluate Assess the effectiveness of the implemented interventions and the system responses to redefine the top priorities and the need for further action.

**Steps:**

1. Implement the strategy and monitor the development of the system
2. Analyse the impacts across sectors and actors
3. Use lessons learned for the next decision-making process

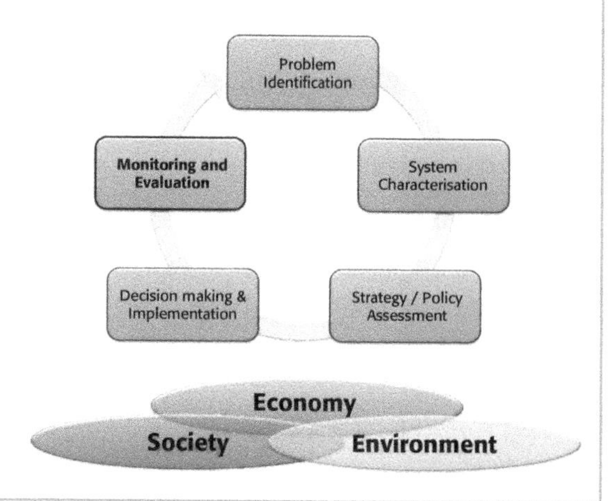


**Tools:**

- Indicators
- Causal loop diagrams
- Simulation

## 7.1 Conceptual mistake #5: Monitoring and evaluation do not affect the decision-making cycle, they only evaluate the system performance

A common mistake in the monitoring and evaluation (M&E) phase is decision makers' reluctance to reframe an unsuccessful approach in light of the lessons learned from the analysis of the system responses. Decision makers harbour the misconception that an M&E exercise is aimed at assessing the system performance and has no bearing on past policies, which means it cannot influence future strategy and policy decisions. In other words, any problem is always a new problem, especially the analysis of possible solutions.

A series of negative consequences result from this erroneous approach to monitoring and evaluation.

First, the informative aspect of M&E is ignored, thus separating a system's performance from the actions and events that triggered change. Such an approach is based on a linear approach to problem solving, which assumes that the same interventions can be successfully adopted to solve similar problems in different contexts (see Section 5.1), regardless of the actual and measured impacts on the system.

Second, if the intervention does not work in the new context, different solutions will be found, but these will not necessarily build on previous experience and would probably not address the main shortcomings of the failed intervention.

> ***An example:*** many companies and non-profit organisations have developed environmentally friendly toilets in recent years. However, these toilets have been either welcomed or strongly rejected, depending on the local context in which they have been introduced. For example, the Stockholm Environment Institute (SEI) designed dry eco-toilets that were successfully introduced in arid areas of the world, where flush toilets connected to municipal sewers were difficult to install. However, the eco-toilet project failed in the Dongsheng district, China, due to insufficient monitoring of development trends in the area. At the beginning of the project, in 2003, the district was lacking adequate sanitation due to water

shortages. However, the rapid development of the area in the following years led to the construction of a 100 km pipeline from the Yellow River to Dongsheng to increase its freshwater supply. As a result, the local population asked for the installation of flush toilets, and became intolerant of the use of eco-toilets, which were perceived as inadequate for their standard of living. In 2009, the government funded the construction of flushing eco-toilets and the dry toilet project was cancelled.

***An example:*** in the 1980s, a company was interested in producing solar cookers for the southern Mexican market. The product had proved successful in other contexts, where it encouraged the use of solar energy and reduced timber extraction activities for cooking purposes. However, an anthropological study warned the company of the possible risks of failure due to cultural patterns in southern Mexico. The study showed that in that part of the country people cook in the morning and the early evening, when solar radiation is limited. Moreover, the cooker could not be used for making tortillas, which are a staple food in the area (Serrie, 1986).

Third, the adoption of a linear and universal approach to M&E also limits the use of essential tools that help with a systemic analysis. In particular, the same small set of indicators (e.g. measuring macro-economic performance by means of GDP) tends to be used to measure strategy and policy effectiveness, which differ in nature and tend to be implemented in different contexts. Similarly, tools such as causal loop diagrams, scenarios and simulations will not benefit the evaluation if indicators tailored to the specific problem (and system) do not inform them.

***An example:*** 'The World', an artificial archipelago of various small islands composed mainly of sand dredged from Dubai's shallow coastal waters, was implemented without adopting a systemic perspective. The project for the realisation of the islands, which were conceived to resemble a miniature of the world's major landmasses, was designed and approved by relying exclusively on the willingness of investors to buy the properties and by assuming continuous growth of the luxury housing market. However, the construction works had to be interrupted because of a number of concurring factors.

First of all, the 2008 financial crisis caused a major drop in property values, forcing the developer to put hundreds of building contracts on hold. Moreover, system responses were largely underestimated during the feasibility assessment. In particular, the islands are sinking as a result of sea currents and the artificial relocation of sands. Further, there is danger of earthquakes due to liquefaction and slope failure, and high vulnerability to extreme natural events. Finally, damage to the local aquatic environment caused by altering the ecosystems might reduce the tourism attractiveness of the area considerably (e.g. lack of water circulation, as experienced in the case of Palm Jumeirah), thereby further lowering the value of the properties. The failure of 'The World' could easily have been avoided if the monitoring and evaluation of the initial project steps had been more carefully analysed.

Fourth, a linear approach to M&E only focuses on the short-term impacts.

Fifth, only the continuous analysis of the system responses allows decision makers to identify early warning signs or gaps. These are two key factors that contribute to the design of additional interventions to address the weaknesses of existing actions, while maintaining their positive impacts.

***An example:*** in October 2010, a major environmental catastrophe happened in the Hungarian town of Ajka, after a dam in the aluminium waste processing plant's reservoir collapsed. Tailings, which is the waste resulting from mining activities, contains toxic substances that can infiltrate rivers and cause enormous damage to human health and the environment. In this case, the disaster approached international levels when the overflowing tailings reached the Danube River, which provides water to about 10 million people across 19 countries. On the other hand, a systemic and continuous approach to monitoring and evaluation could have avoided this catastrophic event. In particular, the personnel responsible for the management of the plant could have detected a number of early warning signals, based on which prompt action should have been taken. First of all, the lack of maintenance had already caused a number of minor leaks and cracks in the dam structure, which were ignored for many years. Also, the streets and natural barriers surrounding the plant would direct any water flow towards the rivers and eventually to the Danube. In addition, weather patterns contributed to the disaster. Before the event, local villagers had expressed concern about the previous season's unusually high rainfall, as well as the strong winds blowing dry tailings mud over farmland. Finally, similar disasters had occurred in other countries of the region in previous years, including the 2009 tailings dam failure of a gold mine in Karamken, Russia, due to prolonged heavy rains.

## 7.2 What to do: Assess the effectiveness of the implemented interventions and the system responses to redefine the top priorities and the need for further action

Once a strategy is being implemented, the system needs to be analysed and compared with the expected results. An understanding of the dynamic system elements—developed through the strategy and policy formulation and assessment process—will certainly help identify gaps and early warning signs in the monitoring phase. Decision makers should thus be able to design complementary actions to reduce any gaps and prevent negative effects from spreading across the system. Therefore, if a problem emerges in the monitoring and evaluation phase (e.g. a target is missed), this information should be used to restart the decision-making process, to analyse the issue and its causes, and to subsequently proceed with system characterisation, policy assessment, decision making, and monitoring and evaluation. It is therefore good practice to start designing a monitoring and evaluation plan at the beginning of the decision-making process.

**What type of simulation?**

Simulation can be utilised to effectively support the strategy/policy assessment (identifying intervention options that will have the desired impact on the system and the right degree of impact) and the evaluation (simulating selected intervention options).

Simulation models are most commonly used when the analysis is done *ex ante*, or before the actual implementation of the interventions.

- **_Ex ante_ modelling can generate 'what if' projections of the expected (and unexpected) impacts of proposed strategy/policy options on a variety of key indicators across sectors. In addition, well-designed models, which merge economic and biophysical variables and sectors, can help with the cost–benefit analysis and the prioritisation of strategy/policy options. The use of structural models that explicitly link interventions with their impacts can generate effective projections of how a certain target could be reached and when.**

On the other hand, an *ex post* analysis is useful to monitor trends and analyse why projections may differ from reality in the months and years after strategy/ policy implementation.

In this context, system response delays should be taken seriously in order to effectively evaluate the suitability of corrective actions. On the one hand, decision makers should refrain from intervening before the effects of the strategy/policy intervention become visible. They should give the system time to react and produce the desired effects. On the other hand, they should not intervene too late: when the initial objectives no longer correspond to the current situation and the delayed system responses can generate unintended consequences.

***An example:*** the use of chemical fertilisers is often identified as a viable intervention to increase soil productivity and food production. However, after an increase in yields in the short term, chemical fertilisers reduce the soil quality, leading to lower yields in the medium and longer term. Conversely, investments in ecological practices and resource efficiency will increase yields in the medium term—after a possible decline in the first two to three years due to the switch to ecological agriculture practices—but will also conserve water and land resources, reduce emissions and allow the ecosystem to recover, which will reverse the environmental deterioration trend observed in a business-as-usual case (Bassi *et al.*, 2011).

*Ex post* modelling can support strategy monitoring and evaluation if the understanding of the relations between the key system variables is improved and by comparing the projected performance with the initial conditions and historical data. Improvements to the model and updated projections allow decision makers to refine the targets and objectives, thus building on the synergies and positive spillovers across sectors.

Three main steps are proposed to carry out monitoring and evaluation effectively:

1. Implement the strategy and monitor the problem
2. Analyse the impacts across sectors and actors
3. Re-start the decision-making process

### 7.2.1 Step 1: Implement the strategy and monitor the development of the system

Once decisions have been made regarding the interventions needed to solve a problem, and the roles and responsibilities of the key stakeholders have been established, the strategies and policies can be implemented. The monitoring phase should start simultaneously with the start of the implementation. The system behaviour should be observed from the launch of corrective measures in order to promptly detect possible synergies and unintended consequences. This will allow decision makers to (re)act in a timely manner.

Monitoring the implementation of interventions is important for a number of reasons:

- It allows decision makers to compare the process performance in comparison with the expectations and evaluate it
- Prompt dissemination of insights from the M&E across all the stakeholders is likely to improve the coordination and alignment of expectations
- Monitoring can be used to improve accountability. This will allow responsibilities to be clearly assigned in order to encourage improved performance and to maximise the results in each phase of the strategy

- Most importantly, monitoring allows a deeper understanding of systemic features, thereby facilitating the identification of early warning signs and worrying trends that might compromise the entire decision-making cycle

***An example:*** the United Nations Development Programme (UNDP) follows precise procedural rules for the monitoring and evaluation of its projects and programmes conducted in 177 countries. In particular, UNDP considers M&E an activity aimed at improving the learning environment within the organisation and at allowing timely changes to be made to projects that fail to deliver the expected outputs on schedule. Relevant data is collected on the ground during the progress towards the expected results. This data is periodically communicated to the agency to which the project reports, as well as to the respective outcome or sectoral monitoring mechanism. Thereafter, the project management evaluates the performance and decides on possible changes to improve the strategy. If revisions are needed, the entire structure of the project has to be reframed, which includes providing result frameworks with new cost estimates, annual targets, etc. Proposed changes are then communicated to the higher-level directors for a final decision (UNDP, 2009). Not only has this created a centralised approach aimed at strengthening accountability at all levels while offering a variety of perspectives from the different relevant departments, but decentralised evaluations have also been created. These evaluations—which the UNDP country offices carry out in collaboration with government partners, other UN agencies and local stakeholders—provide 'more flexibility for strategic choices in evaluation coverage, and a greater range of instruments to discuss with partners' (UNDP, 2011). By combining vertical and horizontal approaches, UNDP promotes an integrated, multi-stakeholder M&E process, which is likely to increase accountability and collaboration, while simultaneously considering different national and local perspectives on development.

Initially, the problem and its key drivers should be monitored to assess the effectiveness of the selected interventions with respect to reaching the desired goals. In this phase, the target indicators should be analysed in relation to the objectives set at the beginning of the decision-making cycle. If the impacts differ from the expectations, an in-depth analysis of the potential roadblocks—known and/or emerging—should be carried out. Effective M&E supports decision makers by determining whether:

- The conditions have changed irrevocably—an urgent change in strategy may be needed
- The system has taken a slightly different direction than expected—which can easily be fixed by means of minor adjustments
- Delayed system responses are part of the normal system behaviour—they prevent sudden and measurable changes within the period of observation

***An example:*** when the managers of Caborca, a medium-sized Mexican company that manufactures cowboy boots and employs 250 people, decided to invest in meeting the requirements of the EU eco-label, they expected to face upfront costs in order to find suppliers of sustainable materials. The strategy, however, proved to be successful, since costs were soon recovered. Continuous monitoring and evaluation of the strategy indicated that a number of factors contributed to the success of the initiative. These include, among others: the lower prices of sustainable materials (they cost an average 8% less), reduced costs related to worker safety due to the phasing out of toxic materials, the minimisation of resource use and waste, increased worker productivity as the result of a safer and healthier working environment, and strengthened customer loyalty due to improved reputation. As a result of this winning strategic approach, the company is planning to launch a new line of ecological boots in Mexico, the United States and Europe (UNEP, 2013).

***An example:*** the launches of Coca-Cola C2 and Zero are an example of a business strategy that failed and the creation of a successful one due to M&E. Coca-Cola C2 targeted male consumers who wanted a low-calorie Coke, but did not like the feminine image associated with Diet Coke. Coca-Cola C2 was launched in 2004 with a US$50 million campaign. The advertisements highlighted that it had half the calories and carbohydrates of a normal Coke. However, the product sales were much lower than expected in the first year and Coca-Cola C2 was eventually withdrawn from the market. Strategy evaluation demonstrated that consumers perceived the benefit of consuming half the calories as an insufficient reason for giving up normal Coke. Lessons learned from this failure led Coca-Cola to commercialise a new product, Coca-Cola Zero, which, like Diet Coke, was free of calories but branded in a way to attract male consumers (Schneider & Hall, 2011).

### 7.2.2 Step 2: Analyse the impacts across sectors and actors

Having assessed the effect of the interventions on the problem and its causes, the M&E phase should proceed with a careful analysis of the interventions' impacts on other sectors and actors that are part of the system. The potential synergies or unintended consequences of planned actions were evaluated in the strategy and policy assessment phase. This was primarily done to ensure a coherent distribution of the costs and benefits across all the relevant stakeholders, but also to work towards improving the entire system's performance, rather than focusing on isolated sectors. In the M&E phase, the expectations are compared with the reality.

Evaluating the impact of interventions across key sectors allows decision makers to assess whether the interventions were formulated and assessed correctly, and whether potential gaps were correctly anticipated. In order to restore the correct, sustainable functioning of the system, the decision-making cycle needs to carefully consider the potential benefits (or harm) that could affect the different sectors. These sectors affect and are affected by not only the original problem, but also the implemented interventions. The assessment should therefore be carried out by means of a system-wide analysis and not a 'siloed' approach that analyses each sector separately.

***An example:*** the damaging effects of industrial pollution on the environment have led many countries to adopt standards and to incentivise the use of cleaner technology in order to reduce emissions from manufacturing processes. Integrated M&E should be conducted and focus on the cross-sectoral reach of these policy interventions, including the impacts on air quality, industrial productivity and employment. Moreover, the effect of green manufacturing should be measured on a variety of related sectors, such as water (reduced contamination of water resources for drinking, household use and sanitation), improved health (due to reduced pollution), fisheries (industries on the coast no longer dump waste in the sea) and waste (industrial waste management).

***An example:*** when evaluating the effects of a stimulus package for eco-tourism, the co-benefits should also be measured in other sectors, such as the provision of local services and the manufacturing of tourism-related goods (e.g. beds, souvenirs, equipment). These, and other activities, impact the well-being of the population through job creation and income, but also through the creation of knowledge and skills. All these factors contribute to poverty reduction and well-being when the poor in the area participate in income-generating activities linked to the eco-tourism industry.

Similarly, the M&E process should aim to ensure that the costs and benefits of interventions are distributed fairly across the actors. A complex problem generally affects a variety of stakeholders, which implies that any proposed solution will affect the same, or even more, stakeholders. The opinions of all these stakeholders need to be considered in the process of evaluating a strategy's success. However, the proposed solutions' impacts also need to be evaluated and analysed regarding their positive or negative effects and goals. A balanced distribution of the responsibilities, costs and benefits will ensure coordination and equilibrium in the system. They should therefore be monitored to make proposed solutions sustainable over time. In this sense, the stakeholders should all collectively carry out M&E activities (UNDP, 2009).

***An example:*** the multi-stakeholder evaluation that a USAID team conducted between October and December 2011 is an example of such an evaluation. The USAID team's evaluation focused on 10 separate projects that different donor agencies funded. Each of the projects was aimed at strengthening the agriculture and livestock value chains in Kenya. The team facilitated key informant interviews, semi-structured interviews and focus group discussions in order to capture information from a variety of stakeholders, including the farmers, the implementing partner staff and other organisations. For example, the Private Sector Development in Agriculture project (PSDA) was evaluated by means of a baseline survey of the farm households, input dealers, service providers and the process owners in eight selected districts. With this approach, the team was able to evaluate how the project costs and benefits were distributed along the agricultural value chain and whether the dual goals of economic growth and poverty reduction were pursued in a balanced way (United States Agency for International Development - USAID, 2012).

***An example:*** EPODE is an international network that aims at reducing childhood obesity. A multi-stakeholder approach is adopted to deliver diet and physical activity-related messages to children and families. The results of the initiative are constantly monitored and evaluated by a Central Coordination Team (CCT), supported by multiple stakeholders from all sectors, including representatives of civil society, corporate sector, NGOs and institutions. The structure ensures that key decision makers and supporters are involved in both community activities and higher-level policy change. This participatory and inclusive decision-making and evaluation system contributed to EPODE's success in preventing obesity in 10 French towns, and to its expansion to other countries, including Belgium, Greece, the Netherlands and Spain (World Economic Forum, 2013).

### 7.2.3 Step 3: Use lessons learned for the next decision-making process

In a best case scenario, and at the end of the M&E phase, the decision-makers should conclude that the problem has been effectively solved, the system performance has improved, the costs and benefits of the strategy/policy are equally distributed between the system's key stakeholders, no side-effects have been created, and new opportunities are arising from unforeseen synergies. In this case, the specific problem-solving process has reached an end. However, because the system evolves constantly, other emerging issues will need to be solved. In this respect, the lessons learned throughout the decision-making process certainly help provide a better analysis of the system's systemic properties and the resulting dynamics.

Conversely, the evaluation of the system's response to the strategy and policy implementation may highlight the failure of the implemented approach. The reasons for failure are many and varied, including incorrect formulation of the strategy and an unpredictable change in the system. Regardless of the specific cause, if the problem is not solved, the decision-making cycle should restart. An improved analysis should be carried out of the causes and effects. A new strategy/policy should be formulated and assessed on the basis of this analysis, after which decisions can be taken regarding the best interventions to implement and monitor.

***An example:*** the government of South Africa recognised that corruption had to be promptly addressed in order to guarantee the basic conditions for investments. The goal was to improve national competitiveness in the global economy. Initially, the problem was addressed from a macro and legal perspective by enacting anti-corruption laws and ratifying international conventions such as the UN Convention Against Corruption and the OECD's Anti-Bribery Convention. However, during the monitoring phase it became clear that legislation was not sufficient to curb corrupt practices in the public and private sectors. By exploring the system's response to these interventions, it was possible to identify the weak links in the chain and reframe the strategic approach. First of all, weak law enforcement was found to be a main cause of policy failure and more precise enforcement mechanisms had to be defined. The focus was on specifically building the capacity of government officials while ensuring appropriate institutional monitoring. The biggest problem encountered during the monitoring phase was the lack of collaboration—primarily due to a lack of trust—between the public and private sectors despite their common interest in fighting corruption. Consequently, the World Economic Forum's Partnering Against Corruption Initiative (PACI) was introduced. PACI is a global, multi-stakeholder, anti-corruption platform seeking to improve dialogue between business and governments in order to have them adopt a systemic approach to tackle corruption and reduce its negative impact on the national economy. As an external initiative and body, PACI has helped overcome the problems encountered during the implementation of the anti-corruption strategy. It has allowed for a more detailed analysis of the causes and impacts of corruption on the different actors and sectors in South Africa, such as in the field of mining and metals. Further, PACI has led the way to a more efficient decision-making process, resulting in the identification of more effective interventions (World Economic Forum, 2009).

Another possible outcome of the M&E process is the recognition that the implementation of interventions gives rise to new issues, regardless of whether the original problem has been solved. New problems are generally the result of the side-effects and unintended consequences of interventions. They may affect other sectors, which siloed choices penalised or damaged, and which are exclusively focused on the sector where the initial problem emerged. In addition, certain actors, who were excluded from the decision-making process and whose interests were poorly represented in the strategy, may be affected. In this case, the strategy and policy should be revised and all the relevant stakeholders should be engaged in the process to select complementary measures that could mitigate and eliminate the anomalies, while maintaining the positive impacts that the previous interventions generated.

### Box 5 Proposed tools: simulations

Simulation refers to the creation of quantitative scenarios, or projections (not accurate predictions), of possible future patterns that may emerge from the system. These scenarios or projections originate from a mathematical simulation model. In the context of this book, simulation exercises are used as dynamic and integrated 'what if' assessments that can create insights into the impacts of strategy/policy implementation.

Simulation can be used to effectively support different phases of the decision-making cycle. In particular:

- Simulations can be used if the analysis is done *ex ante* in the **strategy/policy assessment** and **decision-making** phases. In this context, the use of dynamic simulation models is a fundamental step to test hypotheses, simulate scenarios and learn about the problem's complexity before the actual implementation of the interventions. In particular, *ex ante* modelling can generate 'what if' projections regarding the expected (and unexpected) impacts of the proposed strategy/policy options on a variety of key indicators across the sectors. In addition, well-designed models, which merge the economic and biophysical variables and sectors, can assist with the cost–benefit analysis and prioritise the strategy/policy options. Essentially, the use of structural models that explicitly link interventions and their impacts can generate projections regarding how a certain target can be reached and when.

***An example:*** four industries—iron and steel, aluminium, paper and pulp, and chemicals—account for nearly half of the energy that US manufacturing industries consume and over 10% of the total US energy consumption, making them highly vulnerable to volatile energy prices. Three studies that the National Commission on Energy Policy, the Environmental Defense Fund and the AFL-CIO Working for America Institute (WAI) commissioned between 2008 and 2010 were aimed at creating a dynamic simulation model. With the support of industry association organisations, these studies examined how increased energy prices would affect the competitiveness of these industries in the long term, given the comprehensive and mandatory cap-and-trade climate policy proposals that the US Congress was considering at that time. The studies also examined the industries' capabilities and opportunities to mitigate adverse cost impacts and improve their economic performance under different climate policy scenarios.

In short, the findings strongly suggested that, in the long run, technologies will be available to enable energy-intensive industries to achieve sufficient efficiency gains to offset and manage the additional energy costs arising from a climate policy. However, the use of a system dynamics model identified the need for the analysed industries to design additional measures that could mitigate the cost impacts in the short-to-medium term and could support the creation of policies that would encourage and facilitate

*Continued*

energy-reliant companies to transition to a low-carbon future, while enhancing their competitiveness in global markets. The findings of these studies contributed substantially to the debate on the introduction of climate regulations in the US and abroad (Yudken and Bassi, 2009a, b; Bassi and Yudken, 2009; Bassi *et al.*, 2009c).

***An example:*** Under the leadership of the Ministry of Renewable Energy and Public Utilities of the Republic of Mauritius and with support from UNDP, a system dynamics model was created to support the formulation and evaluation of Mauritius's first longer-term energy policy framework. The goal of this project was to empower the Ministry of Renewable Energy and Public Utilities and the Government of Mauritius with a flexible, integrated, dynamic, user-friendly and exclusive customised simulation model that allows for the evaluation of energy policy proposals so that they can make informed decisions on longer-term policy planning. This model was jointly developed with a team of experts, including representatives from the Ministry of Public Infrastructure, Land Transport and Shipping, the Central Electricity Board (CEB), the Electrical Services Division (ESD), the Mauritius Sugar Industry Research Institute (MSIRI), the Central Statistics Office (CSO), the Maurice Ile Durable (MID) Fund and the University of Technology Mauritius (UTM).

Owing to its flexibility and ease of use, in addition to its integrated and dynamic nature, the Mauritius Model allowed for a cross-sectoral analysis of the impacts of the energy policy provisions, with simulations running from 1990 to 2025. The project included several group modelling sessions and daily exchanges with the with key stakeholders and ended with a two-day workshop and a presentation to the Deputy Prime Minister of the Republic of Mauritius.

The results of the analysis proved to be of considerable value to the Ministry of Renewable Energy and Public Utilities, and led to an update of the longer-term energy policy document, which was later approved. The utilisation of an integrated, cross-sectoral, national development model also served to bring several ministries, the private sector and universities together to jointly analyse the results, as well as the opportunities and challenges arising from the implementation of the energy strategy (Bassi, 2009; Bassi *et al.*, 2009a).

- Simulations can also be used in the **monitoring and evaluation** phase when interventions are being implemented and their performance needs to be monitored. Monitoring the impacts of implemented actions with the help of dynamic models can also help identify potential undesired effects, which will allow decision makers to react in a timely manner, to address the unexpected consequences, and to guide the system through the adaptive process.

Returning to a previous example: after introducing industrial pollution regulations, policy-makers may notice that an increasing number of small and medium enterprises are going bankrupt. The reason for this may be the difficulty that enterprises below a certain size have with surviving in the market. They have to withstand the considerable upfront expenditure required to adapt and optimise production processes in the short term (e.g. the purchase, installation and maintenance of low-energy consumption technologies to replace old machinery). To address this harmful and unexpected consequence, pollution standards and regulations could be complemented with incentives for small and medium enterprises in the form of tax rebates or subsidies, which would partially mitigate the high upfront compliance costs.

***An example:*** the forecasting team at Texas Instruments (TI), a leading DRAM memory chips manufacturer in the 1990s, demonstrated the importance of projections to inform business decisions. The team analysed cause–effect relations in systems, including the expected future capacity and production costs of each manufacturer, the future potential volume of their customers, the market elasticity regarding price changes, and more. Starting with this general analysis, the team was able to explore more detailed strategic questions, such as the potential number of future PC buyers in China, or the impact of expected changes in government subsidies for DRAM manufacturers in various countries. As a result of this exercise, the construction of a new IT manufacturing plant was halted and later discarded on the basis of projections of declining prices and a likely collapse of the DRAM industry. This decision was made at a time when the DRAM market was still expanding and optimism for growth was widespread. However, the decision turned out to be very valid the following year, when a price war started between competitors. This forecasting exercise saved TI and its shareholders billions of dollars (Financial Executives International - FEI, 2001).

## Case study 5 Monitoring managers' commitment: Deutsche Bank

The strategy for tracking managers' commitment at Deutsche Bank (DB) arises from the need to strengthen corporate affiliation in order to avoid losses deriving from unexpected shifts in managerial positions (Büchel and Probst, 2001).

Heinz Fischer, the Head of Corporate Human Resources, realised that the bank needed to better manage the risk of losing its key managers. This need became particularly evident in 2000, when the Asset Management Group of DB lost its biggest customer, New York City Retirement System, to the advantage of Barclays Global Investors and Merill Lynch & Co., following the move of key fund managers from DB to Merill Lynch. It was estimated that this move implied a cost of approximately US$60 billion to DB. As a result, the attachment of top managers to corporate values and goals became a central aspect of human resource management, and it was brought to the top of the corporate agenda. Shortly after, DB calculated that the cost of losing an investment banker exceeded DM636,000. This included the loss of productivity due to the vacancy, expenses for recruiting and the training of new staff members (Büchel and Probst, 2001).

Based on these figures, Fischer drafted a strategy to track managers' commitment using three surveys: 1) a full employee survey, conducted at least once every three years; 2) a corporate identity survey, carried out at least once a year with approximately 4,000 employees; and 3) a 'health check', measuring the current pulse at least three times a year with approximately 1,000 employees. As a result of the study, he developed an indicator to measure managers' commitment, and invited all the heads of section to constantly monitor the indicator in order to identify potentially worrying trends and early warning signs, and take prompt action as needed (Büchel and Probst, 2001).

The strategy proved successful in identifying and measuring the three key components of managers' commitment, namely: engagement, the degree of willingness to perform beyond usually expected levels, and the intention to stay with DB. However, only one head of section acted on the data, initiating a half-day workshop with employees to discuss employee satisfaction, developmental opportunities, job rotation, leadership style and work volume, and identified steps to be taken in order to retain people within that division. The other corporate branches did not find the indicator useful for the purpose of retaining managers, and ignored the findings of the surveys (Büchel and Probst, 2001).

This case study provides relevant insights into the key challenges and opportunities related to monitoring and evaluation. First, it is clear that the monitoring phase is an essential component of the strategy/policy cycle, and should not be underestimated or bypassed. Indeed, DB could have avoided the loss of its biggest client if a proper managers' commitment monitoring strategy were in place. Second, a multi-stakeholder approach is essential to maximise the effectiveness of the monitoring and evaluation phase. In this case, a participatory process for the definition of the strategy would have certainly facilitated the alignment of all corporate branches on how to use the indicators and act on the findings. Finally, monitoring and evaluation becomes ineffective if the information is not used to redefine top priorities and restart the decision-making process.

# 8
# Conclusions

With the ever growing complexity of our world, analysing and solving issues by means of a systemic approach are rapidly becoming a necessity for successful decision makers in the private and public sectors. Strategies and policies need to be designed by means of a multi-stakeholder approach to ensure broad support and effective implementation, as well as to anticipate the emergence of possible side-effects. Only a systemic approach that simultaneously considers social, economic and environmental factors can deliver sustainability.

This book analyses the decision-making process, presents the key conceptual mistakes and solutions for each phase, and introduces specific steps and tools to better identify issues, formulate and assess policies, select and implement interventions, and monitor and evaluate their performance.

Starting with the Bühler case, the book highlights the importance of taking a systemic approach. Specifically, the company re-evaluated its failed strategy by identifying problems from an integrated perspective. To do so, Bühler widened the range of indicators used to formulate and assess the new strategy, and explicitly considered the following elements of the system: technology, suppliers, product, government, marketing and customers. This consequently led to the inclusion of a broad range of stakeholders in the company's strategy development process.

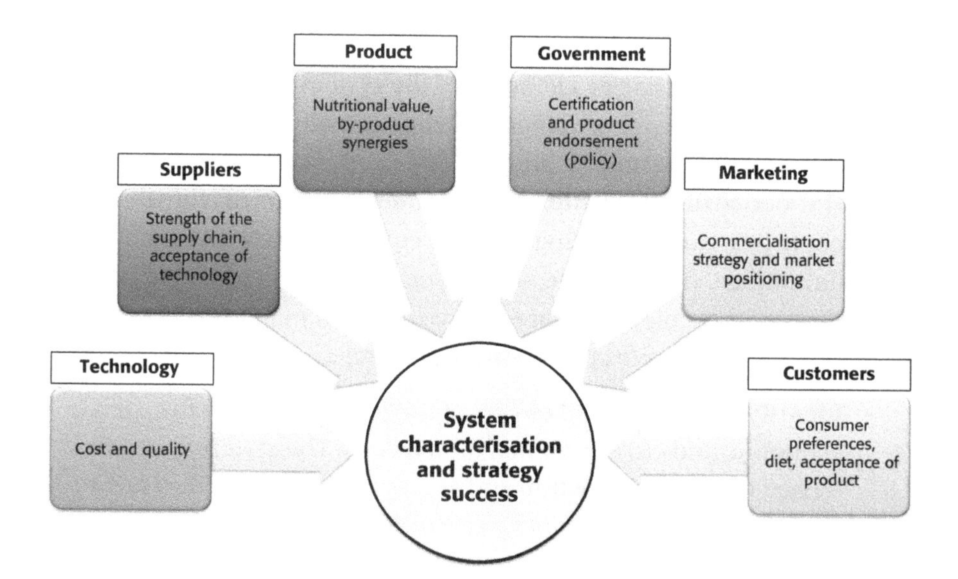


FIGURE 13 The main elements of the system analysed by Bühler in the creation of a new strategy

Having identified the key stakeholders and the main elements of the system, both internal and external to the company, the analysis of the network started with the use of influence tables. The relationships among the stakeholders, the interdependencies among the main elements of the system, and the enabling conditions were identified. These included, among others, the importance of the government for certifying the product, without which market access would have proven impossible.

Further, success was achieved through the identification of the main levers for interventions, also using causal diagrams. In fact, each system was built on specific factors that could be controlled and others that could not. In this context, it was crucial for Bühler to use controllable elements to influence the system and, in the case of certification, find a domestic partner for the analysis of NutriRice™ to ensure certification.

Scenarios were analysed to assess various business and marketing strategies (e.g. creation of its capacity, or a strategic partnership with the local millers) and to inform the decision making. This was done by taking a systemic approach and considering the potential side-effects that could emerge within the system.

Once the strategy was finalised, implementation, monitoring and evaluation followed. All the key elements of the system were monitored and evaluated against key performance indicators. Finally, the replicability was assessed to introduce new technology and products in the same market, and to introduce the production and commercialisation of NutriRice™ in other countries, carefully taking the local context into consideration.

To summarise, this example and the main contents of this book provide a few general guidelines for effective decision making using systems thinking to solve complex problems, which are:

- **Identify the causes and effects of the problem** across the social, economic and environmental dimensions. The system is characterised by feedbacks within and across sectors, which may create synergies or cause the emergence of side-effects
- **Use a multi-stakeholder approach** to account for a variety of points of view and to incorporate as much cross-sectoral knowledge as possible in the analysis
- **Evaluate the impacts across sectors and find a balanced strategy** aimed at improving the performance of the entire system, rather than at maximising selected indicators at the expense of others
- **Evaluate the impacts across actors and find an inclusive strategy** that will allocate costs consistently and distribute the benefits equitably across the key actors in the system
- **Think long term and prioritise resilience** because success often relates to resilience in the light of unforeseen events, a focus on increasing a system's capacity to absorb change and adapt to it with clear, long-term goals
- **Monitor the performance of the system** to learn the many ways systems respond to strategy/policy implementation, and to improve decision making by incrementally addressing the cause of success and failure of implementation

The challenges ahead are substantial, but the method and tools presented in this book are available to turn them into effective and sustainable opportunities.

# Annex: Overview of key tools

This annex introduces selected tools that can support each step of the integrated decision-making process. These are certainly not the only tools available; the selection was based on their proven effectiveness and the ease with which they can be implemented.

The proposed tools combine speed of execution/analysis with a solid method and also complement one another. This allows decision makers to effectively improve their understanding of the challenge, to identify and evaluate options, as well as monitor their performance systemically and 'organically'. While complexity increases each step of the way—from identifying the issue to evaluating the strategy/policy—the use of the simple tools presented below makes tackling the issue manageable and increasingly intuitive, allowing decision makers to **'grow' and shape winning strategies from complexity**.

The format for the presentation of the tools focuses on the following items:

- Definition
- Key features (strengths and weaknesses)
- Associated decision-making steps
- Possible combination with other tools
- Implementation steps

The tools selected have different characteristics, but complement one another within the decision-making cycle. For instance, indicators and influence tables are 'static'; they only provide information on the past and current system behaviour. Causal loop diagrams (CLDs) add a dynamic component, but also make use of indicators and influence tables. CLDs can be considered the bridge between the static and dynamic tools. The latter are scenarios and simulation aimed at understanding future trends that structural system changes drive.

Overall, the tools proposed cover a broad spectrum and allow the static representation of the system to become dynamic by improving decision-makers' understanding of its underlying functioning mechanisms and allowing them to identify and evaluate effective intervention options.

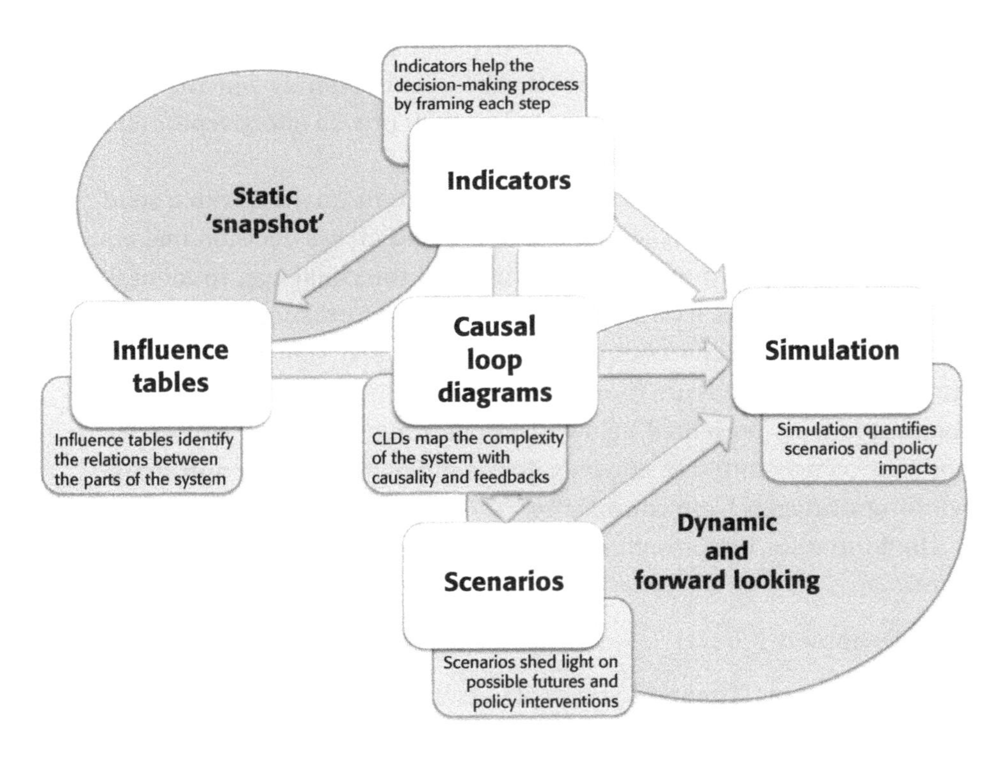


FIGURE 14 Overview of the tools for decision making addressed throughout this book

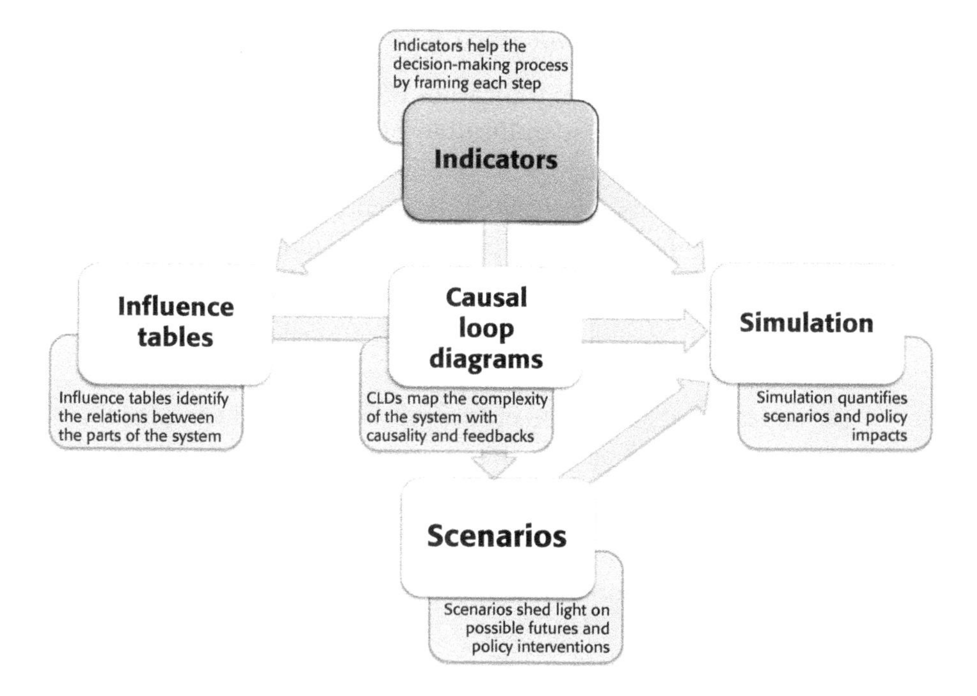


# A1 Indicators

## A1.1 Definition

Indicators are a crucial tool to identify and prioritise issues. They provide clear information on the historical and current state of the system and highlight trends that can shed light on causality to better detect the key drivers of the problem.

Within the integrated decision-making process, indicators can be used to: 1) identify issues and their primary causes; 2) carry out a cost–benefit analysis to evaluate the intervention options; and 3) support the integrated monitoring and evaluation of the strategy/policy impacts (UNEP, 2012b).

Indicators can be divided into three main categories:

- **Problem identification indicators (Phase 1)**. These indicators seek to facilitate the identification of issues. Regardless of the nature of the problem to be solved (environmental, social or economic), indicators are selected that can best identify the problem and its (often many and varied) causes and effects

  ***An example:*** indicators can shed light on the observed climatic changes—such as on the causality between energy consumption, deforestation, GHG emissions and temperature changes—with all the impacts that these have on climate variability and agricultural productivity, among others.

- **Strategy/policy and assessment indicators (Phase 3)**. This group of indicators assesses the potential costs and performance of various intervention options that could be utilised to solve the issue.

  In the context of climate change, these indicators can effectively support the preparation of a cost–benefit analysis. This analysis will evaluate the net investment required for climate mitigation and adaptation that will result in improved climate resilience, in investments being allocated across the key sectors and in employment being created. The net cost/saving of the intervention should be estimated by comparing upfront investment and the economic savings (i.e. the avoided costs and/or added benefits) accrued over time.

- **Strategy/policy evaluation indicators (Phase 5)**. These indicators aim at assessing the success of strategy/policy interventions. Impacts have to be calculated using an integrated approach to strategy/policy evaluation, which includes the development of: 1) human well-being, especially if public policies are involved; and 2) other operations in the business if the private sector is involved

  Improved resilience to climate change can, for example, be measured in terms of environmental benefits (e.g. the sustainable use of natural resources), economic gains (e.g. reduced costs due to climate-related damage) and social advancement (e.g. job creation, poverty alleviation and social inclusiveness).

Table 6 provides an illustration of how indicators can be used to support the decision-making process. Indicators have an important role in helping to identify problems and interventions and to measure their impacts by using the key steps listed in the table and presented in more detail in the following sections.

| Indicators for problem identification | Indicators for strategy/ policy assessment | Indicators of strategy/ policy evaluation |
|---|---|---|
| • Identify the potential worrying trends | • Identify the strategy/policy objectives | • Measure the strategy/policy impacts on the issue |
| • Assess the problem and how it relates to the environment | • Identify the intervention options | • Measure the strategy/policy impacts across the key (business) sectors |
| • Analyse the causes of the issue | • Analyse the costs and benefits of each option | • Analyse the impacts on society, the economy and the natural environment |
| • Analyse how the issue impacts society, the economy and the natural environment | • Select the best options and formulate the related strategy/policy interventions | |

TABLE 6 Use of indicators to support key decision-making steps

## A1.2 Key features

| Strengths | Weaknesses |
|---|---|
| • Support problem identification<br>• Allow for the quantification of problems and identification of their causes and effects<br>• Facilitate the objective evaluation of the intervention options<br>• Enable quantitative monitoring and evaluation of the implemented actions | • Data not always reliable and coherent across sectors<br>• Need to be combined with other tools for systemic analysis (e.g. causal maps, scenarios, simulations, etc.) to effectively support decision making |

TABLE 7 Strengths and weaknesses of indicators

### A1.2.1 Strengths

Indicators are instruments that help decision makers identify and prioritise problems, set the agenda for interventions, as well monitor and evaluate investments. Indicators provide quantitative information that allows decision makers to identify and analyse trends, as well as evaluate the system's current state.

In order to elaborate effective strategy/policy plans to solve problems, the issues have to be correctly identified through a careful analysis of their causes and effects; this analysis has to be undertaken across all the sectors that influence or are influenced by them. Only when the causes of the problem have been correctly identified, can policies that will have a lasting positive impact be designed and the emergence of side-effects be avoided.

If the indicators have been properly selected and analysed, decision-makers will be able to objectively evaluate the considered intervention options' adequacy and performance regarding the problem, the target and their broader social, economic and environmental impacts.

Once a strategy/policy has been designed and implemented, its economic, social and environmental impacts have to be monitored and evaluated. Strategy/policy impact indicators are essential to evaluate the performance of policies and to contribute to the next policy-making round (starting, again, with identifying the issue).

### A1.2.2 Weaknesses

Indicators provide information on the system history as they allow an analysis of the trends. However, the data shown is not always reliable. The quality of the data is key and its coherence across the indicators is of paramount importance. Consequently, indicators cannot be used in isolation and their coherence across the sectors needs to be compared.

Further, indicators cannot contribute extensively to the analysis of complex problems unless a systemic approach is simultaneously used. The analysis of the historical trends and causality helps identify issues. However, a systemic approach (which influence tables and causal diagrams highlight) is essential to identify emerging trends and analyse the continuously evolving underlying system mechanisms. Among others, distinguishing between stock and flows is crucial to evaluate the data quality, the coherence of the chosen indicators, and the interdependencies in the system.

## A1.3 Associated decision-making steps

The initial stage of the integrated decision-making cycle consists of identifying key issues that prevent us from reaching desired goals and targets. Indicators help evaluate the state of the system on the basis of its historical and current performance.

The second and third stages of the integrated decision-making cycle consist of defining the strategy/policy goals in order to then formulate and assess the interventions. While indicators for problem identification help frame the issue, indicators for strategy/policy assessment help design and evaluate solutions *ex ante* by identifying which of the key variables should be influenced to achieve the stated goals with the selected intervention options.

The fifth stage of the integrated decision-making cycle consists of monitoring and evaluating the strategy/policy impacts. While indicators for problem identification help frame the issue and indicators for strategy/policy help design solutions, indicators for strategy/policy impact help estimate the broader impact of the chosen interventions.

Figure 15 indicates the specific use of indicators in the integrated decision-making process. The following paragraphs present the three categories of proposed indicators in more detail.

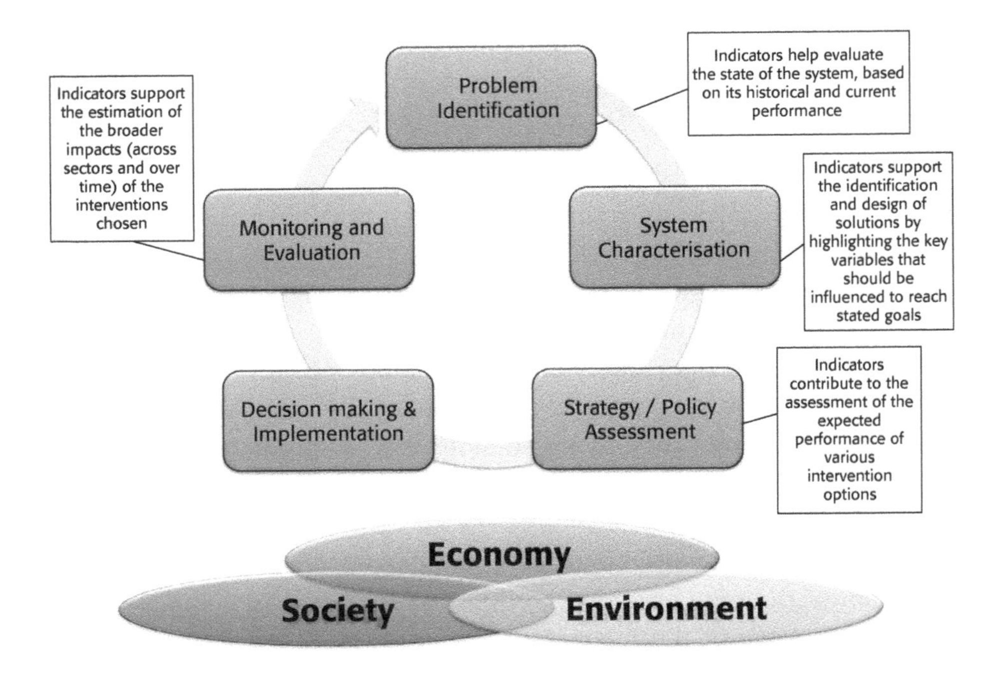


FIGURE 15 Indicators to inform the integrated decision-making process

### A1.3.1 Problem identification indicators

An agenda is a list of issues or problems (including potential opportunities, which may be missed without interventions) to which decision-makers should pay serious attention at any given time.

The agenda for a country might include rising food prices, air pollution and illegal immigration, while a private company might consider a declining profit margin or market share, as well as a lack of technological competitiveness and innovation.

Of all the conceivable issues to which decision makers could be paying attention, they can effectively only pay attention to a sub-group of prioritised issues at a time. This focus also avoids the destabilisation of the systems if too many investments (changes) are simultaneously made. Indicators are therefore a crucial tool to identify and prioritise issues.

### A1.3.2 Strategy/policy formulation and assessment indicators

Strategy/policy formulation is a process that generates intervention options in response to a problem on the agenda. In this process, decision makers frame, refine and formalise strategy/policy options to prepare the ground for the decision-making stage.

In this regard, indicators can be useful at different levels of the strategy/policy formulation, namely: to define the objectives, evaluate the intervention options and costs, to provide a cost–benefit analysis and to formulate the intervention options.

- **Strategy/policy objectives indicators.** Indicators are useful tools to identify specific strategy/policy options and targets

***An example:*** in the specific case of climate change mitigation policy, the indicators of GHG emissions and their key drivers (e.g. fossil fuel consumption and deforestation) can be used to set a specific emissions reduction target (e.g. the percentage reduction of GHG compared with the 1990 level or the business-as-usual (BAU) scenario).

- **Intervention options and costs indicators.** Various intervention options are available with which to achieve specific strategy/policy objectives, while various other indicators can be used to evaluate them

Climate change mitigation policy interventions include the use of demand and supply options in the energy sector. On the demand side, the growth of energy consumption could be curbed through energy conservation (e.g. behavioural change) or through energy efficiency (e.g. installing more efficient appliances and/or light bulbs). On the supply side, renewable energy could be used for power generation, or more efficient thermal power plants could be built. Indicators can be used to assess the current situation (e.g. how much renewable energy is used to generate power and at what cost) and the extent to which these interventions can effectively curb the growth of energy consumption, as well as the associated costs.

- **Cost–benefit and multi-criteria analysis indicators.** A cost–benefit analysis and possibly also a multi-criteria analysis are necessary to evaluate the net investment required, as well as the additional pros and cons—per actor or economy-wide—of the analysed intervention. This analysis should compare the investment and avoided costs or the added benefits, generally, depending on the issue to be solved.

Returning to the climate change mitigation policy example: the adoption of energy-efficient technology requires upfront investments, but will reduce energy consumption and expenditure, while possibly creating new jobs. The added benefits and avoided costs can be compared with the indicators of the sector's historical and current performance to assess whether the investment can be sustained and how the economic burden can be allocated across the main actors affected by the intervention (e.g. public versus private investment).

- **Strategy/policy assessment indicators**. Once the investments required have been estimated, the intervention options can be designed to allocate the costs across the key economic actors. The options include capital investment, as well as public sector incentives and regulations. Indicators can be used to evaluate the best strategy or policy option (or mix of policies) that will not affect a single actor (e.g. households or the private sector) excessively

If no incentives are offered, regulations (e.g. public mandates) imply that, to comply with the new law, only the private sector is responsible for the investment (e.g. a mandate or regulation that renewable energy should form part of the generated power, or that introduces emission standards). Indicators of private sector investment could be monitored to evaluate whether the new policy would require a considerable reallocation of resources (possibly reducing employment and/or requiring external financing), which could lead to broader negative economic impacts. In this case, incentives could be introduced for which the government is responsible—utilising indicators regarding the annual deficit and debt—and which should be monitored.

To conclude, the range of indicators required to effectively support strategy/policy formulation is relatively broad and depends on the specific issue to be analysed.

### A1.3.3 Strategy/policy evaluation indicators

Integrated strategy/policy evaluation refers to the effort of monitoring and determining how an intervention has fared during its implementation. It examines the means employed, the objectives served and the effects caused in practice.

To determine the actual effects of an intervention, its objectives and its broader impacts need to be considered. In this respect, three main categories of indicators can support integrated strategy/policy evaluation: impacts on the sector, the broader socioeconomic consequences and the broader environmental consequences.

#### Sectoral impacts (investments, performance and jobs)

This group of indicators has an economic focus and captures interventions' indirect impacts. These indicators aim at estimating the business's and the related economic sectors' performance, which includes the investment leveraged and the employment generated and/or substituted.

With regard to jobs, the following possible development should be considered (UNEP *et al.*, 2008):

- Additional jobs are created—for example, through the manufacturing of pollution control devices that are added to existing production equipment
- Some employment is substituted—for example, when shifting from fossil fuels to renewables, or from using landfills and waste incineration to recycling
- Certain jobs will be eliminated without direct replacement—for example, when packaging materials are discouraged or banned and their production is discontinued
- Many existing jobs will simply be transformed and redefined when day-to-day skill sets, work methods and profiles are adjusted to new and, possibly, more advanced functions

Indicators should be developed to capture the four elements above, including the nature of the employment created, lost and transformed.

### Indicators of progress and well-being

This group of indicators refers to the overall measures of economic progress and human well-being, including dimensions such as poverty alleviation, equity, social inclusiveness, overall well-being, capital resources and inclusive wealth. These indicators generally apply to public policy interventions, but several private sector interventions also impact the development of a local community, either through the products and services offered, or by establishing production facilities.

The direct and indirect impacts of interventions should be considered, including improved access to energy, water and nutrition, as well as reduced health problems and mortality due to the upgrade to modern forms of energy (through infrastructure improvements and higher income) and higher quality education and business-related skills.

Besides the human development indicator (HDI) and the gender development indicator (GDI), several other aggregate indicators, such as the Millennium Development Goals (MDGs), could improve transitioning and developing countries.

| Problem | Problem identification indicators | Policy interventions | Strategy/policy formulation and assessment indicators | Strategy/policy evaluation indicators |
|---|---|---|---|---|
| Deforestation | *Identify potential worrying trends* | *Payments for ecosystem services (PES)* | *Identify policy objectives* | *Measure the policy impacts regarding the environmental issue* |
| | Deforestation (ha/year) | | Reduced deforestation (e.g. 50% reduction by 2030) | Reduced deforestation (e.g. 50% reduction by 2030) |
| | Annual harvest of wood products ($m^3$/year) | *Agroforestry incentives* | Decrease in the use of wood energy for cooking (tons of oil equivalent/year) | Decrease in the use of wood energy for cooking (tons of oil equivalent/year) |
| | | *Timber certification* | Increased revenues from eco-tourism ($/year; % of GDP) | Increased revenues from eco-tourism ($/year; % of GDP) |
| | *Assess the problem and its relation to the natural environment* | | *Identify intervention options* | *Measure policy impacts across sectors* |
| | Forest land cover (ha) | | PES: funding transferred ($/year and/or $/ha) | Increased revenues from river transport activities ($/year) |
| | Degraded forest land (ha or % of forest land) | | Agroforestry development: investment per ha ($/ha/year) | Increased water supply (L/year) |
| | | | Timber certification: activities certified (#/year and output) | Reduced flood risk ($/year; % of GDP) |

TABLE 8 Example of indicators for policy interventions in the case of deforestation (continued over)

| Problem | Problem identification indicators | Policy interventions | Strategy/policy formulation and assessment indicators | Strategy/policy evaluation indicators |
|---|---|---|---|---|
| | *Analyse the causes of the environmental issue*<br>Agriculture land (ha)<br>Rainfall (mm/year)<br>Population (people) | | *Analyse costs and benefits of each option*<br>Cost of incentives ($/ha)<br>Forestry employment (people)<br>Forest goods production ($/ha or % change) | *Analyse impacts on the overall well-being of the population*<br>Employment and income generation (people/year, $/year)<br>Carbon stored in biomass (ton)<br>Infrastructure investment avoided (water, road) ($/year) |
| | *Analyse how the issue impacts society, the economy and the environment*<br>Income of forest communities ($/year)<br>Freshwater supply (L/year)<br>Eco-tourism (visits/year; $/year) | | *Select the best options and formulate related policy interventions*<br>Employment created per $ invested (people/$)<br>Biodiversity increase per $ invested (GEF index for biodiversity)<br>Income creation for rural communities ($/year) | |

**TABLE 8** (from previous page)

## A1.4 Possible combination with other tools

As indicated above, indicators support all the decision-making process steps. Indicators therefore also complement the other tools proposed in this book.

The approach proposed consists of utilising indicators to identify the problem by means of influence tables and CLDs, which allow for a more systemic analysis of the causal relations to identify the key causes and effects of the problem to be solved. Indicators are therefore used to set targets and identify possible strategy/policy options that could mobilise the required investment. CLDs allow the effectiveness of interventions to be tested by identifying their key entry point, as well as their direct and indirect impacts throughout the system. Finally, indicators are used to select key variables in order to create scenarios and simulation models.

## A1.5 Implementation steps

The methodology proposed focuses on four main steps to identify indicators to be used to identify the issue. The DPSIR framework (UNEP, 2010a) could help identify the indicators as it can identify the drivers (D), pressures (P), state (S), impacts (I) and responses (R):

- **Identify potential worrying trends**. The initial step to identify an issue that might constitute a threat is to analyse its historical trend. This can be done by using historical quantitative data and qualitative information if reliable statistics are not yet available
- **Assess the problem and its relation to the environment**. Once a trend has been identified and defined as worrisome, indicators need to be selected to evaluate whether there is a link between the problem and its environment
- **Analyse the causes of the issue**. Once the problem has been detected by analysing and comparing the indicators of economic, social and environmental performance, and the relation between the problem and the environment has been identified, the causes of the environmental indicators' underperforming trend can be identified and analysed. The pressures and driving forces (causes) should be clearly separated from the symptoms (the environmental impacts and the state of the environment)

- **Analyse how the issue impacts society, the economy and the environment**. In the first three steps, we analysed the trends and identified the key causes of the issue to ensure that the problem is properly addressed and that the decision makers receive the identified information. This step extends the analysis to the impacts that the underperforming environmental trend may have on the other indicators. This allows us to identify additional problems that can be solved by addressing this underlying environmental cause and to further prioritise the issues and the need for intervention

The key steps to identify and analyse the indicators for strategy/policy formulation and assessment concern the availability and choice of intervention options. These are also called the 'enabling conditions' and consist of investments, as well as national regulations, policies, subsidies and incentives.

The methodology proposed focuses on the four main steps to identify the indicators for strategy/policy formulation and assessment:

- **Identify strategy/policy objectives**. The identification of strategy/policy objectives is based on the outcomes of the issue identification phase and precedes the identification and choice of interventions. The targets have to be specific, quantitative in nature, and should be achieved within a specified time. They also have to be ambitious, yet achievable through effective interventions

- **Identify intervention options**. After the targets have been established, various intervention options can be considered to achieve them and various indicators can be used to evaluate the usefulness and effectiveness of these options

- **Analyse the costs and benefits of each option**. A cost–benefit analysis is necessary to identify the best and the poorest of the options in terms of the costs and benefits. Simply put, this analysis should generally compare the investment and the avoided costs, or added benefits, depending on the issue to be solved

- **Select the best options and formulate related strategy/policy interventions**. Once the various strategy/policy options available to solve the problem and to achieve the stated targets have been analysed with the help of the indicators of the costs and benefits, the best combination of interventions needs to be chosen. Selecting the strategy/policy options should be based on three criteria: the cost should be shared fairly, the problem should be solved effectively, and cross-sectoral double and triple dividend opportunities should be taken into consideration

The approach used to identify the strategy/policy impact indicators covers a broader set of consequences, which are of a social, economic and an environmental nature. These indicators include information on the state of the environment, which is directly related to the environmental issues and target indicators, as well as the indicators of the sectoral performance and socioeconomic progress, such as employment and well-being.

The methodology proposed focuses on three main steps to identify the indicators for strategy/policy monitoring and evaluation:

- **Measure the strategy/policy impacts regarding the issue**. When monitoring and evaluating the impacts of interventions, the indicators for issue identification should be analysed to test the actual effect of the implemented interventions
- **Measure strategy/policy impacts across key sectors (business)**. After having measured the effectiveness of the strategy/policy intervention in solving the problem at hand, the cross-sectoral impacts should also be measured. Given the social, economic and environmental indicators' high degree of interdependence, each intervention implemented in one sector is also likely to have impacts (either positive or negative) on the other sectors
- **Analyse the impacts on society, the economy and the natural environment (well-being, performance)**. To ensure sustainability, the economic, social and environmental impact indicators need to be monitored in the strategy/policy evaluation phase. Several indicators can be used to estimate the impact of interventions on well-being, including employment and income generation, and the sector performance, as well as access to resources (e.g. energy, water, sanitation) and health (e.g. harmful chemicals in water, people hospitalised due to air pollution).

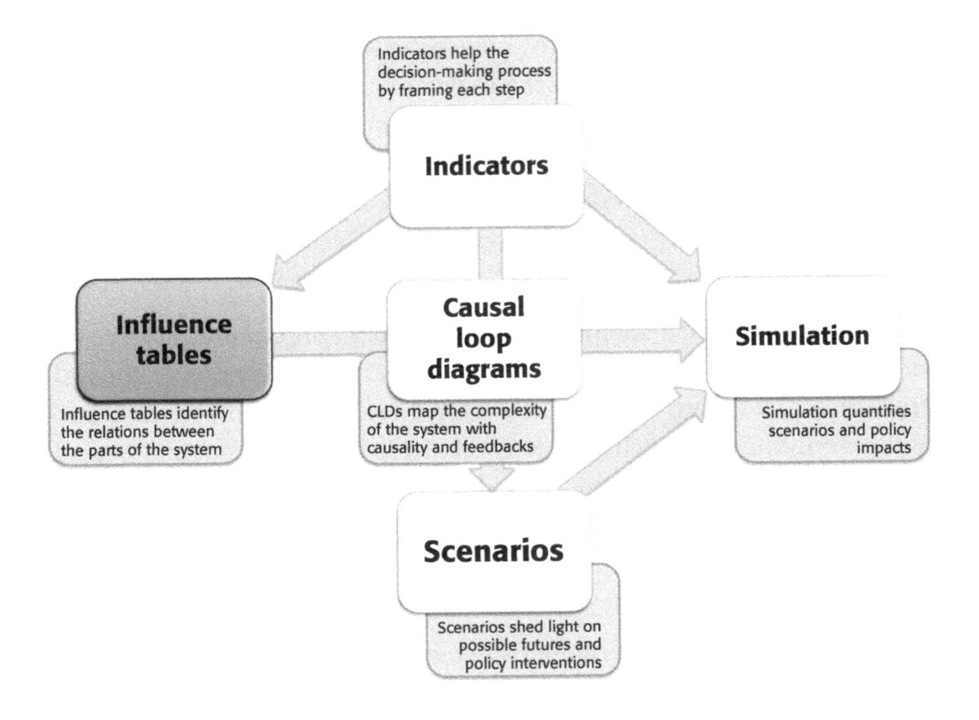


# A2 Influence tables

## A2.1 Definition

An influence table is a matrix that relates each of the (selected) system variables to the others and indicates how strong the causal link (or influence) between these variables is (based on Vester, 1978; Vester and Hesler, 1982; Probst and Gomez, 1989, 1995).

The strength is generally defined using a scale from 0 (no influence) to 3 or 5 (strong influence). The blanks along the diagonal of the matrix show that the individual variables cannot influence themselves directly.

Influence tables are useful to identify the causal relations among key variables in the system, as well as the strength of these relations. This tool allows decision makers to visually identify the key drivers of change, as well as the key effects of changes in key variables in the system.

In order to facilitate this process, the **active sum** (AS) and **passive sum** (PS) are calculated. A **quotient** (Q = AS/PS *100) and a **product** (P = AS*PS) are also estimated for each variable. These are used to characterise the variables as follows:

- **Active variable** (highest Q) influences the others strongly, and is little influenced by others
- **Passive variable** (lowest Q) influences the others very little, and is a much influenced variable
- **Critical variable** (highest P) influences other variables strongly, and it is also strongly impacted by other variables in the system
- **Inert variable** (lowest P) has a limited influence on other variables, and is only slightly influenced by them

Characterising key variables as active, passive, critical or inert provides a good starting point for decision makers to identify the main levers for intervention. When coupled with the delay of each specific intervention option (concerning implementation and/or system response), influence tables effectively inform the creation of causal loop diagrams.

Table 9 is an example of an influence table, highlighting the strength of the causality between selected key variables in the system. In this case, a standard example of a commercial activity is presented.

| Effect of → on ↓ | Wages | Sales mix | Prices | Sales | Competitive position | Profits | Staff bonus | Active sum AS | Quotient Q (AS/PS*100) |
|---|---|---|---|---|---|---|---|---|---|
| Wages | – | 0 | 0 | 2 | 1 | 3 | 2 | 8 | 133 |
| Sales mix | 1 | – | 0 | 3 | 2 | 2 | 0 | 8 | 200 |
| Prices | 0 | 2 | – | 3 | 1 | 3 | 0 | 9 | 900 |
| Sales | 1 | 1 | 1 | – | 2 | 3 | 2 | 10 | 77 |
| Competitive position | 1 | 0 | 0 | 3 | – | 2 | 0 | 6 | 75 |
| Profits | 2 | 1 | 0 | 0 | 2 | – | 3 | 8 | 57 |
| Staff bonus | 1 | 0 | 0 | 2 | 0 | 1 | – | 4 | 57 |
| Passive sum PS | 6 | 4 | 1 | 13 | 8 | 14 | 7 | | |
| Product P (AS*PS) | 48 | 32 | 9 | 130 | 48 | 112 | 28 | | |

TABLE 9 Example of an influence table for a commercial activity

## A2.2 Key features

| Strengths | Weaknesses |
|---|---|
| • Support the identification of causal relations and their strength<br>• Identify key drivers of change in the system<br>• Allow entry points for action to emerge from the analysis<br>• Can be used intuitively, straightforward interpretation of results<br>• Favour multi-stakeholder engagement in strategy/policy formulation | • Only present the relationship between two variables at a time. Multiple influence can only be inferred<br>• Analysis is static<br>• Cannot fully support strategy/policy formulation and assessment; need to be linked to causal diagrams |

**TABLE 10 Strengths and weaknesses of influence tables**

### A2.2.1 Strengths

Influence tables allow for identifying and representing the strength of the relation between key variables (i.e. indicators) in the system. As a result, in the context of the system's functioning mechanisms, they provide an intuitive representation of the interrelations characterising the problem to be solved.

The main variables identified as active, passive, critical or inert also provide decision makers with information on possible entry points to influence the behaviour of the system and solve the problem.

Influence tables simplify complexity and support dialogue among stakeholders during the investigation of the key drivers of change and their relations with the system.

### A2.2.2 Weaknesses

Influence tables only present the relations between two variables at a time. Simultaneous influence can be inferred from the analysis of the *product* and *quotient*, but these calculations only provide system-wide estimates of the relevance of each variable.

These tables are static. Delays can be coupled to the estimation of the strength of the relations, but influence tables are 'accounting frameworks' rather than dynamic instruments to evaluate the system's possible future behavioural paths. As such, they inform decision making by identifying entry points for action, but do not effectively support strategy/policy formulation and assessment.

## A2.3 Associated decision-making steps

Influence tables primarily support the decision-making process in its early phases.

This tool can be used in the problem identification phase, during which causal relations are identified and their strength is assessed. The result of this exercise influences strategy/policy formulation directly by highlighting key entry points for intervention.

Further, influence tables support the monitoring and evaluation stage, during which system-wide impacts can be assessed using the *product* and *quotient* to explain the strategy/policy-induced changes in the system behaviour.

In line with the other tools proposed in this book, influence tables favour the adoption of a multi-stakeholder approach, leverage the use of indicators and provide crucial inputs to create a causal loop diagram.

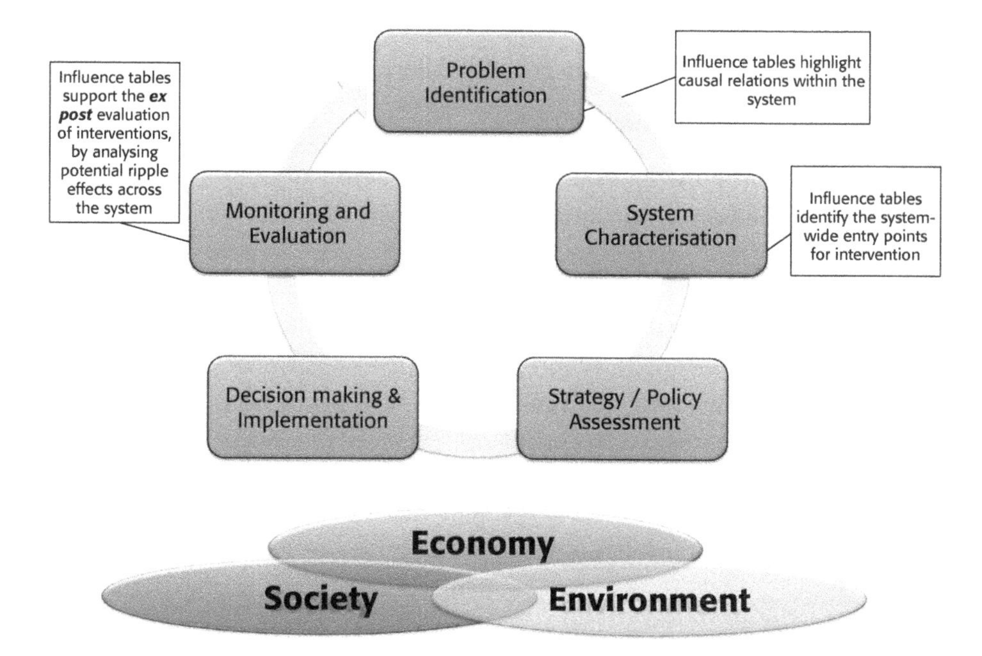


FIGURE 16 Influence tables to support the decision-making process

## A2.4 Possible combination with other tools

Influence tables make use of indicators that can identify, estimate and analyse the strength of causal relations in the system. Problem identification indicators, as well as selected strategy/policy formulation indicators, can be utilised to create influence tables.

Influence tables are crucial for the creation of coherent and useful causal diagrams. They are an intuitive and solid intermediary step between indicators and causal loop diagrams, which greatly enhances the decision maker's understanding of the issue/opportunity and cooperation among the team members and the stakeholders.

Influence tables are the first tool to 'see' and explore systems in the decision-making process and do so by means of a matrix, with which most decision makers are normally familiar (rather than the learning curve required to correctly create and use causal diagrams).

## A2.5 Implementation steps

The creation of influence tables is straightforward and requires four main steps:

- The identification of the key variables in the system
- The preparation of a matrix
- The estimation of the strength of the relation among variables (to be carried out with a multi-stakeholder approach)
- The calculation of the sum, product and quotient

With these steps completed, the analysis of the table can start. In this regard, the first step is to identify active, passive, critical or inert variables. Following this step, entry points for intervention are selected, which are then further assessed by means of the causal diagram.

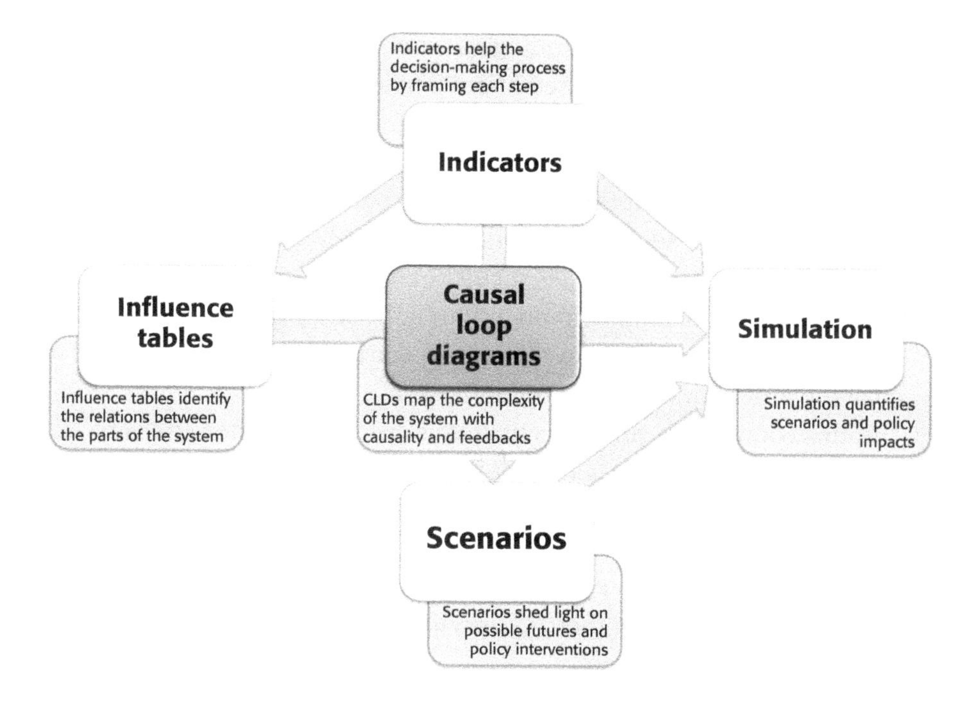


## A3 Causal loop diagrams

### A3.1 Definition

A causal loop diagram (CLD) is a map of the system analysed or, better, a way to explore and represent the interconnections between the key indicators in the analysed sector or system.

A more accurate definition is that a CLD is an integrated map (because it represents different system dimensions) of the dynamic interplay (because it explores the circular relations or feedbacks) between the key elements—the main indicators—that constitute a given system.

By highlighting the drivers and impacts of the issue to be addressed and by mapping the causal relationships between the key indicators, CLDs support a systemic decision-making process aimed at designing solutions that last.

The creation of a CLD has several purposes. First, it combines the team's ideas, knowledge and opinions. Second, it highlights the boundaries of the analysis. Third, it allows all the stakeholders to achieve basic-to-advanced knowledge of the analysed issues' systemic properties.

Having a shared understanding is crucial for solving problems that influence several sectors or areas of influence (e.g. departments in a multinational company), which are normal in complex systems. Since the process involves broad stakeholder participation, all the parties involved need a shared understanding of the factors that generate the problem and those that could lead to a solution, to effectively implement successful private–public partnerships. As such, the solution should not be imposed on the system, but should emerge from it. In other words, interventions should be designed to make the system start working in our favour, to solve the problem, rather than generating it.

In this context, the role of feedbacks is crucial. It is often the very system we have created that generates the problem, due to external interference or to a faulty design, which shows its limitations as the system grows in size and complexity. In other words, the causes of a problem are often found within the feedback structures of the system. The indicators are not sufficient to identify these causes and explain the events that led to the creation of the problem.

We are too often prone to analyse the current state of the system, or to extend our investigation to a linear chain of causes and effects, which does not link back to itself, thus limiting our understanding of open loops and linear thinking.

Causal loop diagrams include variables and arrows (called causal links), with the latter linking the variables together with a sign (either + or –) on each link, indicating a positive or negative causal relation (see Table 11):

- A causal link from variable A to variable B is positive if a change in A produces a change in B in the same direction
- A causal link from variable A to variable B is negative if a change in A produces a change in B in the opposite direction

| Variable A | Variable B | Sign |
|---|---|---|
| ↑ | ↑ | + |
| ↓ | ↓ | + |
| ↑ | ↓ | – |
| ↓ | ↑ | – |

TABLE 11 Causal relations and polarity

Circular causal relations between variables form causal, or feedback, loops.

> The energy policy that has been in place in Saudi Arabia in recent years is a good example of a feedback loop that can be found in real life. In order to distribute the exceptional profits of the country's oil exports, the government decided to subsidise the domestic gasoline prices to a greater extent when world oil prices increased. This mechanism helped maintain the country's social cohesion. On the other hand, this intervention generated a series of side-effects: the lower the domestic price of gasoline, the higher the domestic consumption; when domestic consumption increases, all else being equal, exports, as well as profits, decrease. In order to mitigate this negative effect, Saudi Aramco, the national oil company of Saudi Arabia, had to increase its domestic refining capacity to avoid paying a premium price to foreign refiners, which normally refine exported crude oil, and to maximise the domestic production's profitability.

This example shows a negative feedback loop: the current high profits lead to a decrease in future profits due to the increasing domestic demand. Such loops tend towards a goal or equilibrium, balancing the forces in the system (Forrester 1961).

A feedback can also be positive when an intervention in the system triggers other changes that amplify the effect of that intervention, thus reinforcing it (Forrester 1961).

> This happens with an oil field's production before it reaches a plateau phase: the higher the investment in the production capacity, the higher the production. Likewise, the higher the production, the higher the revenues and, therefore, the investments in the production capacity and production. Further, in the plateau and decline phases of the production, the balancing loops—driven by depletion—will dominate.

## A3.2 Key features

| Strengths | Weaknesses |
|---|---|
| • Facilitate a multi-stakeholder approach to problem solving<br>• Help highlight the causal relations between the indicators<br>• Support the analysis of the system behaviour and its reaction to external interventions | • Effectiveness is strictly linked to the process quality<br>• Wrong or partial CLDs may lead to ineffective (or even harmful) interventions<br>• Best used if combined with quantitative tools (e.g. simulation models) |

TABLE 12 Strengths and weaknesses of CLDs

### A3.2.1 Strengths

CLDs highlight the drivers and impacts of the issue to be addressed and map the causal relations between the key indicators.

By explicitly identifying the feedback loops, CLDs shed light on the main internal mechanisms that led to the problem and also allow projections to be made regarding the system's possible future trajectories in reaction to any implemented decision.

They help identify entry points for interventions and evaluate their effectiveness, as well as the synergies and potential side-effects.

They help avoid 'blaming' for failure and promote the identification of systemic solutions by clarifying that the causes of a problem are found within the feedback structures of the system and are not due to uncontrollable external events. An external event is not even a problem as such, but the way the system reacts to this event is.

### A3.2.2 Weaknesses

The effectiveness of a CLD is directly related to the quality of the work and the knowledge that goes into developing the diagram. Multi-stakeholder perspectives should be incorporated and cross-sectoral knowledge is essential to correctly identify the causes of the problem and design effective interventions.

The boundaries of the system and the relationships between the key variables have to be correctly identified. Errors in creating the diagram may lead to the implementation of policies that do not generate the desired effects, and may even backfire.

The estimation of the strength of causal relations, even if these are correctly identified, cannot be guaranteed as the causal diagram is a qualitative tool. It is therefore advisable to use a causal diagram together with a similar integrated and dynamic causal descriptive mathematical simulation model.

## A3.3 Associated decision-making steps

CLDs support the decision-making process in several ways and provide valuable input during each step.

More specifically, in the agenda-setting phase, CLDs allow for identifying the causal chain that identifies the problem to be solved. The CLD can therefore show decision makers problems that may have been overlooked. We too often focus our attention on an event (i.e. the manifestation of a problem) rather than on the problem. By explicitly showing the causal relations and feedback loops, a CLD allows the mechanisms that led to the creation of the problem to be identified, which leads to a far more accurate problem identification effort.

In the strategy/policy formulation and assessment phases, CLDs allow for identifying the key entry point for interventions. With CLDs, it is possible to identify the weakest link in the system and to target key feedback loops that (when strengthened or neutralised) will generate positive change. Further, CLDs allow decision makers to follow the causal chain and to identify all the changes generated in the system. This also allows them to identify the system responses to the implemented interventions.

In the strategy/policy evaluation phase, CLDs help evaluate the interventions' performance. This takes place on two levels: 1) short vs. long-term impacts and responses; and 2) direct and indirect impacts and responses. The system reacts to the interventions implemented, possibly generating synergies, but perhaps also creating side-effects and elements of strategy/policy resistance, which make the intervention ineffective.

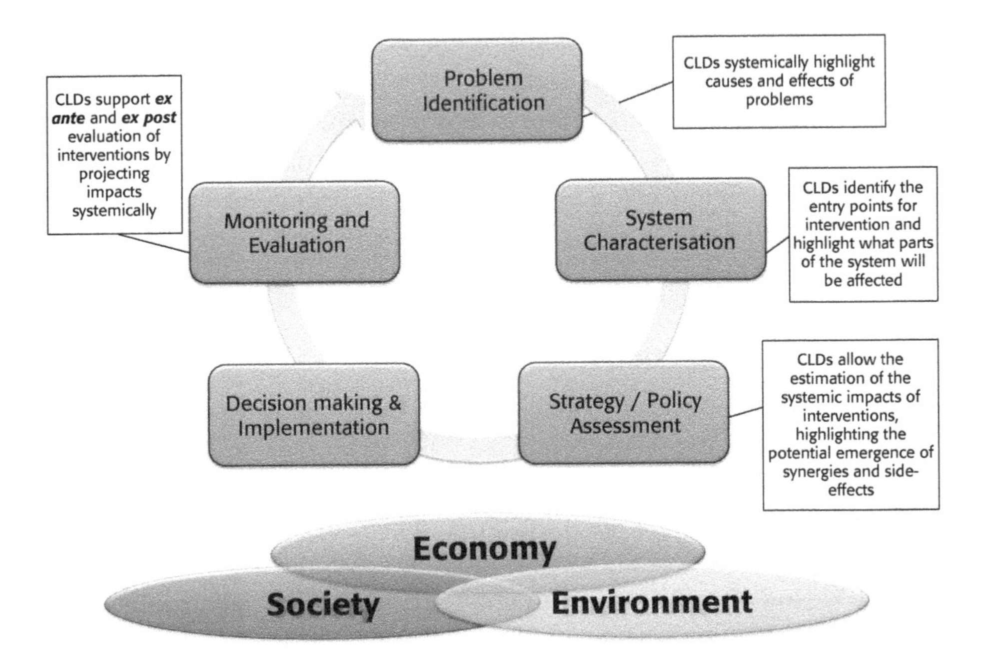


FIGURE 17 CLDs to support the decision-making process

## A3.4 Possible combination with other tools

Indicators are an integral part of CLDs, as they are directly used to create the map of the system. Further, CLDs allow the future trajectories of key indicators to be evaluated, making the analysis of future scenarios dynamic.

CLDs complement influence tables by adding 'dynamics', which allows the patterns of behaviour, as well as the interconnections across key variables, to be analysed. On the other hand, a CLD does not show the strength of these relations across the key indicators and agents, which are shown (and qualitatively estimated) in the influence tables.

CLDs support the creation and analysis of scenarios, because they highlight endogenous and exogenous drivers, as well as the way they respond to external interventions. Consequently, CLDs help make the scenario analysis more concrete, ground it in the actual mechanisms of the analysed system and add an element of response from the system (i.e. systems do learn and change state when interventions are implemented). Owing to the high dynamic complexity that has to be managed, the system response would be very difficult to analyse coherently if the scenario were not mapped and visible.

CLDs are also necessary for the correct development and creation of integrated and dynamic simulation models. The creation of a CLD is actually the second step of the modelling process, after the 'issue identification'. Therefore, CLDs complement a simulation model and define the scenarios that have to be simulated.

## A3.5 Implementation steps

As mentioned above, a CLD will only be as good as the knowledge and work put into it. On the other hand, a few additional steps have to be followed to design a useful and effective causal diagram.

The basic knowledge needed to build a CLD includes the polarity concept (i.e. the sign of the causal relation between two variables, whether positive or negative) and the feedback concept (reinforcing or balancing), as mentioned above. The following are the practical steps that should be followed:

- Start with the key indicator identified as representing the problem and add it to your diagram (which is blank at this stage)
- Add the causes of the problem, one by one, linking them to the first variable considered and determine the polarity of the causal relation
- Continue identifying and adding the cause of the cause, and so forth

In the process, the diagram will grow and other variables will influence some of the variables identified as causes of the problem. These circular relations are the feedback loops (representing closed-loop thinking), which are also the key functioning mechanisms of the analysed system. Thinking in terms of feedbacks is crucial in the development of CLDs and requires a multi-stakeholder approach.

More specifically, the following recommendations should be followed to create a good causal diagram (Sterman, 2000):

- Use nouns or noun phrases to represent the elements rather than verbs. That is, the links (arrows) represent the actions in a causal loop diagram and not the elements. For example, use 'cost' and not 'increasing cost' as an element

- Generally it is clearer if you use an element name in a positive sense. For example, use 'growth' rather than 'contraction'
- A difference between the actual and perceived states of a process can often be important to explain patterns of behaviour. In many cases, there is a lag (delay) before the actual state is perceived. For example, when there is a change in actual product quality, it usually takes a while before customers perceive this change
- There are often differences between short-term and long-term consequences of actions and these may need to be distinguished with different loops
- Keep the diagram as simple as possible, subject to the earlier points. The purpose of the diagram is not to describe every detail of the management process, or the system, but to show those aspects of the feedback structure that lead to the observed problem. In other words: model the problem, not the system

Finally, once the creation of the diagram is complete, the analysis can begin. Normally the starting point is the first variable added to the diagram, or the key problem to be solved. It is good practice to 'read' the diagram to understand the extent to which simultaneous factors influence the causes of the problem. Further, reading the diagram helps check its consistency and validity and also identifies the overall system pattern and the main feedback loops responsible for it.

There are a few methods to determine whether a feedback loop reinforces or balances. The two most commonly used are:

- **Reading the CLD**. Starting with the assumption that the first variable in the loop will increase when the loop is followed: 1) we end up with the same result as in the initial assumption (i.e. that the variable increases) and the feedback loop reinforces; 2) we end up contradicting the initial assumption (i.e. that the variable decreases) and the feedback loop is balanced, or opposes change
- **Counting plus and minus signs**. 1) Reinforcing loops have an even number of negative links (zero is also even); 2) balancing loops have an uneven number of negative links

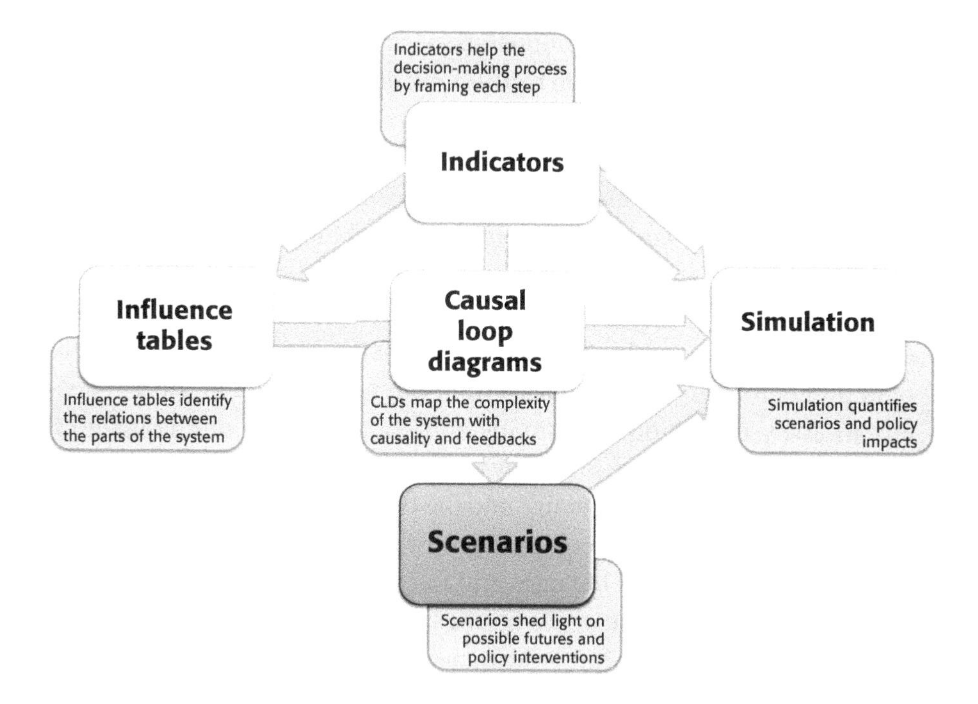


## A4 Scenarios

### A4.1 Definition

Scenarios are expectations about possible future events used to analyse potential responses to these new and upcoming developments. Consequently, scenario analysis is a speculative exercise in which several future development alternatives are designed, agreed upon, explained and analysed for discussion on what may cause them and the consequences these future paths may have on our system (e.g. a country or a business).

Scenario analysis is designed to improve decision making, allowing it to embrace uncertainty and risk, and allowing the analysis of patterns and paths, as well as strategic responses. Scenario analysis can also be used to explore unexpected events, increasing the general readiness to expect unforeseen external impacts on the system.

Both the private and public sector have used scenario analysis over the last few decades to manage risk and develop robust strategic plans in the face of an uncertain future.

In the private sector, Shell pioneered scenarios and due to the effectiveness of, and the publicity given to, the work done, they have been increasingly used to manage large capital investments and change corporate strategy. Scenario analysis is now a standard tool for planning activities that span the medium and long term (hence, they can be successfully applied to capital investment where the capital life time is an average of 25–30 years).

Scenarios have also helped public policy formulation and governments around the world are increasingly using them. Initially used to plan population growth, scenarios were extensively used in the last decade to analyse the potential trends and consequences of changing oil prices. Currently, they are primarily used to analyse the impact of human activity leading to climate change and the resulting consequences regarding natural disasters and economic growth (e.g. reduction in agriculture yield due to temperature changes and rainfall variability). Further, using scenarios to highlight the opportunities, risks and trade-offs in national policy debates is a niche but growing use, primarily due to international organisations' work.

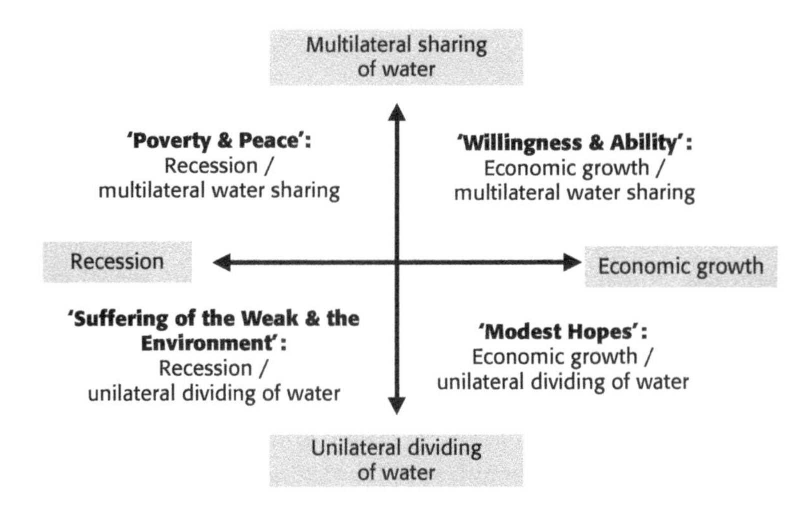


FIGURE 18 GLOWA Jordan River scenarios of regional development under global change

## A4.2 Key features

| Strengths | Weaknesses |
|---|---|
| • Promote innovative thinking about possible future behavioural paths of the systems<br>• Improve multi-stakeholder, cross-sectoral risk management<br>• Support the monitoring and identification of potential future challenges resulting from implemented actions | • Inadequate for short-term planning<br>• Require continuous revision and control by a specialised team<br>• Require a disciplined monitoring, analysis and communication process<br>• Ineffective if not integrated into the decision-making process |

TABLE 13 Strengths and weaknesses of scenarios

### A4.2.1 Strengths

- **Innovative thinking.** Scenario analysis does not rely on history to generalise possible future paths. In other words, it does not extrapolate the past and does not expect past patterns (or the key drivers of past behaviour) to still be valid in the future. In this respect, scenario analysis promotes innovative thinking, examines the dynamic properties of the system systemically and does not exclude any potential system response a priori

- **Improving understanding of risk management.** Scenarios help decision makers identify elements of uncertainty (see Fig. 19). Often scenario planning focuses on the analysis of the external events' impact on our system, forcing decision makers to elaborate concrete contingency plans and exit strategies. The analysis of these plans' effectiveness, which a diversified team or group of stakeholders usually carries out, allows decision makers to simultaneously consider the political, social, economic and environmental dimensions of the system

- **Monitoring progress and anticipating future changes.** Scenario analysis creates expectations about future paths. As a result, indicators can be identified to monitor whether any of the observed paths is actually taking shape. This allows decision makers to effectively, rigorously and systemically monitor the system performance as a whole and to anticipate potential emerging challenges and opportunities

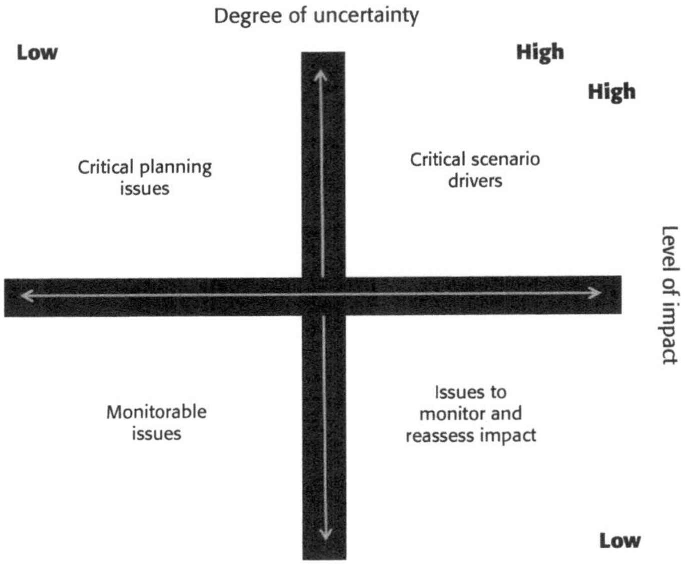

Degree of uncertainty
Low
High
High
Critical planning issues
Critical scenario drivers
Level of impact
Monitorable issues
Issues to monitor and reassess impact
Low

FIGURE 19 Example of an impact/uncertainty matrix

A4.2.2 Weaknesses

- **Long-term and high-level strategic decision making**. Scenarios are designed to support specific activities. Methodologies and models are designed and built for a specific purpose and should never be applied out of context. Scenario planning is useful for longer-term analysis, often for strategic decisions that are not taken daily. The continuous repetition of the exercise may simply add confusion, as the methodology does not apply well to technical decisions that have to be implemented in the short term
- **Scenarios are a process, not a product**. To be effective, they need to be used and refined over time to stay up to date with the system's continuous evolution, which is due to new external events, but also the growing nature of any socioeconomic system. This requires sustained commitment and a skilled and dedicated scenario team

The main challenges in the process can be summarised as follows:

- Failure to integrate scenarios in the decision-making process, making them a useless exercise from a decision-making perspective
- Failure to develop a clear map of the future with monitorable indicators. If not monitored, scenarios cannot be used to systemically anticipate challenges and opportunities
- Failure to correctly analyse and capture the properties of the system. The process requires discipline and attention, and a diverse set of inputs and points of view. If the scenario exercise does not identify the system's true key drivers, it will lead to inaccurate results and actions
- Failure to communicate the results effectively. Scenarios are mostly qualitative exercises. If the results are not communicated properly, they will not be used for decision making

## A4.3 Associated decision-making steps

Scenarios particularly support the first phases of the integrated decision-making process: agenda setting and strategy/policy formulation. Scenarios can also support strategy/policy evaluation, but since the process is primarily oriented towards the analysis of emerging future trends, it is better coupled with activities that look forward rather than backward.

More specifically, scenarios are elaborated to explore possible future paths and trends, analyse patterns of behaviour and drivers of change in order to ultimately elaborate interventions that will allow our system to adapt (e.g. to cope with challenges and/or profit from new opportunities).

Consequently, scenarios are useful in the agenda-setting process: if an emerging trend is observed, and new issues have to be faced, the company's agenda or the government's agenda will change to incorporate new concerns.

Further, since scenario analysis includes the exploration of the system responses to external or internal (man-made, such as policies) events, it can also contribute to the strategy/policy formulation and assessment stage. At this level, the main question is: *what are the implications of each scenario for the business/country?* This question points to the decisions to be made and possible management changes that will have to be implemented to cope with the system's expected evolution under each scenario. Despite the qualitative nature of scenarios, concrete interventions should be proposed to allow indicators to be identified and trends to be monitored in order to anticipate future challenges and opportunities. Considerations along this line include (Maack, 2001):

- Which interventions, if any, are suggested under all the scenarios? If the same intervention is proposed for all the scenarios, it is very likely already needed and should therefore be included in a follow-up policy package or strategic plan
- Which of the suggested interventions, if any, are diametrically opposed to the current ones? The analysis of this aspect serves to reduce risks related to the current strategy, even if the expected scenario does not become reality
- Which interventions, if any, are expected to strengthen the current strategy? This question helps anticipate opportunities with the existing interventions, which allows synergies to be found and the effectiveness to be maximised

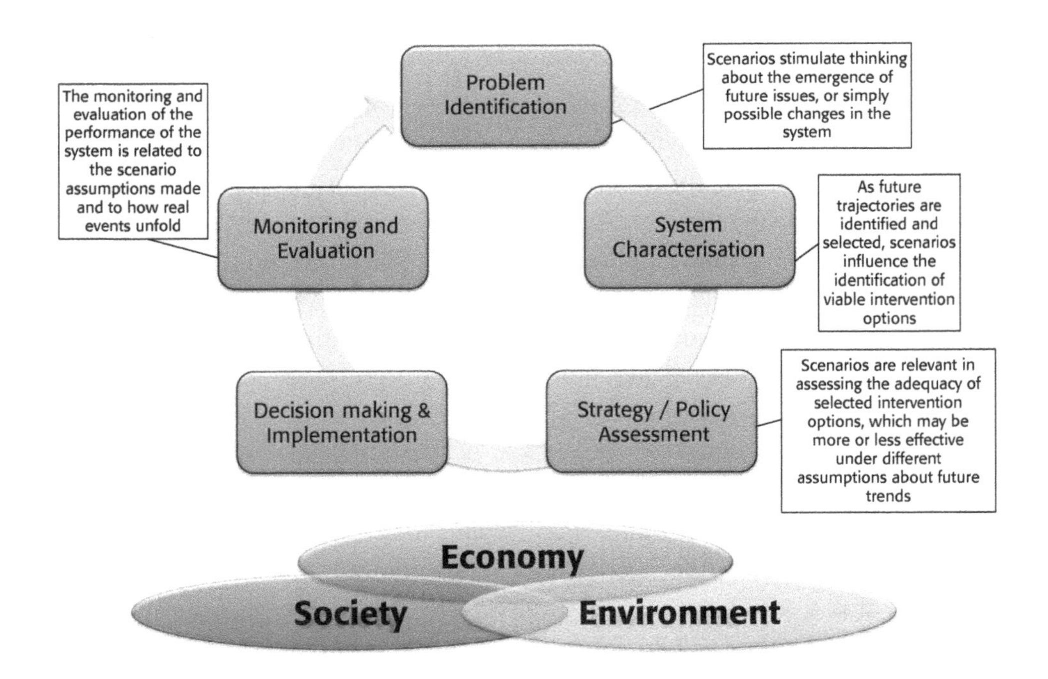


FIGURE 20 Scenarios to inform the integrated decision-making process

## A4.4 Possible combination with other tools

Scenario planning is useful, even when used in isolation, to explore emerging patterns, regardless of whether external events or internal system developments trigger them. Developing scenarios requires several tools presented in this book and also contributes to them in various ways.

Scenarios explore the structure of the system to provide an understanding of how it may react to imposed and/or emerging changes. While this is done qualitatively and in the form of assumptions, CLDs help formalise the mapping of the system by clearly identifying the causal relations and the feedbacks responsible for internally generated changes in the system, as well as the system responses to external changes. Consequently, the use of CLDs in the scenario-developing process adds rigour and objectivity, ensuring a shared understanding of the system's main driving mechanisms.

At the same time, CLDs can make use of the results of scenario planning. Several scenarios can be analysed with a single CLD and often requires no change to the variables and/or causal relations. Scenarios can actually identify the main external factors that will gain relevance in the future and which should be added to a CLD for an analysis of the responses that the system will generate.

Along the same line, simulation could also make use of the complementarity of CLDs and scenarios, turning the qualitative analysis into a quantitative assessment.

Finally, scenarios make use of, and contribute to, the use of indicators. The selected indicators mean that scenarios are built on observations of reality. Furthermore, by looking forward in time, new scenarios may highlight that a new set of indicators should be monitored to anticipate upcoming challenges and opportunities.

## A4.5 Implementation steps

Scenario planning is commonly implemented in many ways and in various different styles. The main steps of the process are presented in Figure 21 and recognise that several variations could be implemented to allow a better adaptation to the specific context and the complexity of the issue to be solved. Other tools' complementarity with scenarios is clearly indicated in order to generate a solid analysis by means of a coherent and comprehensive process.

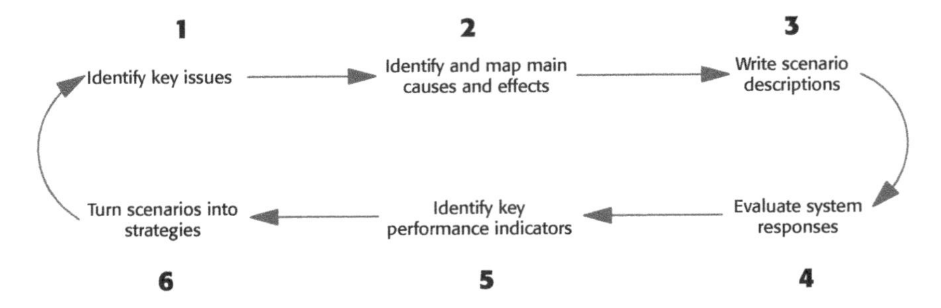


FIGURE 21 Diagram of the scenario process

Scenario analysis involves **developing scenarios** (steps 1–3 below), and **analysing interventions and system responses** (steps 4–6 below).

1. **Identify the issue to be analysed**. Scenarios are best suited to look at the future through the lens of a specific issue, which external events (e.g. a public policy or a natural disaster) could generate. This step can be carried out in conjunction with the use of indicators (for issue identification) and CLDs
2. **Identify the key causes and main drivers of the system**. These are the social, economic, environmental, political and technological factors that are most relevant for the issue and are responsible for generating the system's observed and future patterns of behaviour. This step can be carried out in conjunction with the use of influence tables and CLDs
3. **Write scenario descriptions, fully explaining the underlying drivers (e.g. feedbacks) of each scenario**. These are the stories that explain how driving forces interact and what effects they have on the system and the analysed problem. This step can be carried out in conjunction with the use of indicators (for issue identification) and CLDs (reading a CLD already helps create a storyline for the scenario)
4. **Evaluate the system responses of each scenario**. Each scenario is likely to trigger different feedbacks and response mechanisms.

These should be analysed to better understand the role of each feedback (i.e. which ones reinforce and balance and, of these, which dominate the system at any given point in time?). Intervention options are also analysed at this stage. This step can be carried out in conjunction with the use of indicators (strategy/policy formulation), CLDs and simulation

5. **Identify the main indicators to be monitored to anticipate upcoming challenges and opportunities**. Indicators help decision makers monitor the system and identify the correct entry point for and timing of intervention. This step can be carried out in conjunction with the use of indicators (for strategy/policy evaluation), CLDs and simulation (for *ex ante* analysis)
6. **Turn scenarios into strategies**. Once scenarios have been built and refined, they should be translated into actionable interventions, presented clearly and logically, and implemented as needed. This step can be carried out in conjunction with the use of indicators (at all levels), CLDs and simulation

Ideally, the process presented above should be implemented over a series of workshops (possibly during a day session as well if the team is well built and focused) and should lead to the creation of scenarios and actions that are plausible, consistent, relevant, creative and challenging.

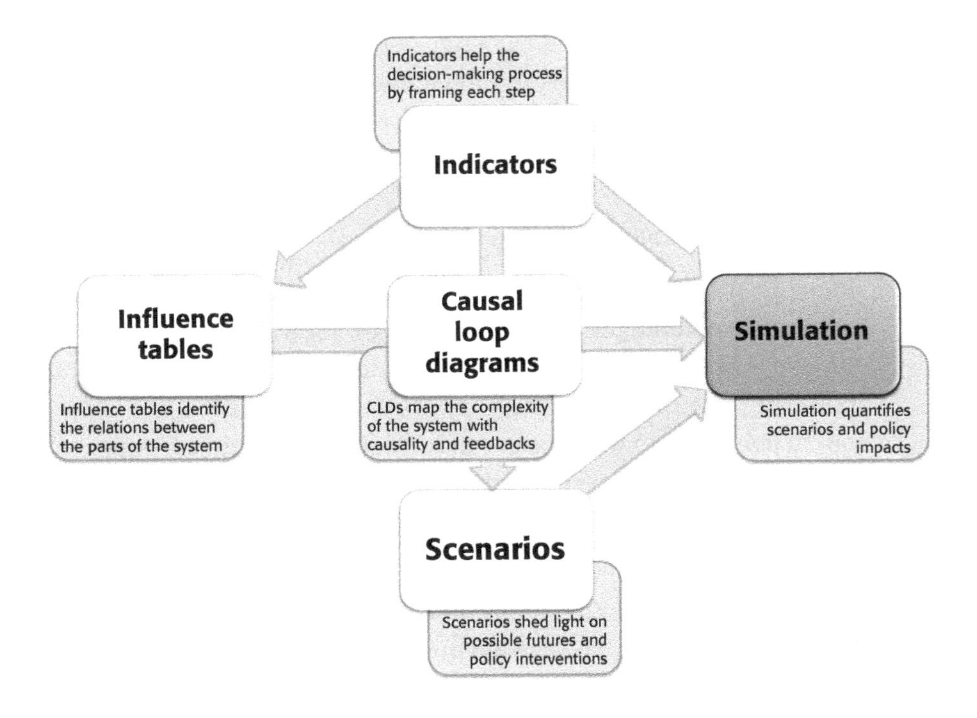


# A5 Simulation

## A5.1 Definition

Simulation refers to the creation of quantitative scenarios, or projections (not accurate predictions) of possible future patterns emerging from the system. Simulations originate from a mathematical simulation model.

In the context of this book, simulation exercises are dynamic and integrated 'what if' assessments that can create insights into strategy/policy implementation's impacts.

- **Dynamic**. Indicating that feedback loops, delays and non-linearity—the three core elements that underlie the characteristics of real systems—are explicitly taken into account and represented in the model
- **Integrated**. Indicating that the simulation model has to include cross-sectoral indicators and incorporate the economic and biophysical dimensions of the problem and analysed system

- **What if**. Indicating that models should not tell us what to do, nor how to optimise our system. Instead, they should provide an evaluation of the impacts of the decisions considered for implementation and should test whether these can effectively solve the problem

On the other hand, mainstream modelling approaches are neither dynamic nor integrated, indicating that there is a clear disconnect between our 'systemic' thinking and available models. This disconnect is due to the emergence of problems, such as climate change, which introduced the concept of feedbacks to the global scene (IPCC, 2007). To ensure that simulation supports the integrated decision-making process effectively, the emphasis should be on modelling methodologies and tools that allow a full representation of the causes and mechanisms responsible for the problems' emergence, as well as for the solution. This also implies that models should take the stocks and flows into account, as it has lately become clear that stocks (e.g. debt) are responsible for the performance of flows (e.g. GDP).

***An example:*** the key drivers of a green economy, as defined by UNEP, are the stocks and flows of natural resources and those of capital and labour, which are important in any long-term economic model. Stocks are accumulations of inflows and outflows (like forests are the accumulation of reforestation and deforestation). In the GER analysis, capital and labour are needed to develop and process natural resource stocks. Thus, three key factors transform natural resources into economic value added: the availability of capital (which accumulates with investments and declines with depreciation), labour (which follows the world demographic evolution, especially the age structure and labour force participation rates) and stocks of natural resources (which accumulate with natural growth—when renewable—and decline with harvest or extraction). The availability of fish and forest for the fishery and forestry sectors, as well as the availability of fossil fuels to power the capital needed to catch fish and harvest forests, are examples of the direct impact of stocks of natural resources on GDP. Other natural resources and resource-efficiency factors affecting GDP include water stress and waste recycling and re-use, as well as energy prices, all of which are endogenously determined. The analysis carried out in the GER focuses on the transition towards a green economy, which is characterised by high resource efficiency and low-carbon intensity. The analysis assesses the need for a short to medium-term transition and evaluates the impacts of a longer-term green economic development. Longer-term sustainable growth is related to the sustainable management of natural resources, such as water, land and fossil fuels. Increasing such resources' efficiency of use and curbing waste would reduce the decline of stocks, or even support their growth in certain cases. In this respect, understanding the relationship between stocks and flows is crucial (e.g. the concentration of emissions in the atmosphere may keep increasing, even if yearly emissions are kept constant or decline. Carbon concentration will only decline if yearly emissions are below, for example, the natural sequestration capacity of forests and land) (UNEP, 2011a).

## A5.2 Key features

| **Strengths** | **Weaknesses** |
|---|---|
| • Facilitate the estimation of strategy/policy impacts | • Require structural and behavioural validation |
| • Support the understanding of complex system dynamics | • Inadequate for accurate 'predictions' |
| • Provide immediate feedback | • Provide a simplified version of reality |
| • Offer quantified, realistic insights | |

TABLE 14 Strengths and weaknesses of simulation

The presentation of simulation's strengths and weaknesses focuses on system dynamics models, the methodology suggested to address complex problems, which are cross-sectoral (affecting, and being impacted by, economic and biophysical indicators and feedbacks) and evolve continuously.

#### A5.2.1 Strengths

- **Estimation of strategy/policy impacts across sectors**. The simulation of scenarios with quantitative models allows decision-makers to evaluate the impact of selected interventions within and across sectors
- **Debunking complexity**. The simulation of causal descriptive models helps simplify and understand the complexity and evaluate the short vs. longer-term advantages and disadvantages of the analysed interventions
- **Prompt feedback**. Currently, the simulation of a scenario with causally descriptive models takes no longer than a couple of seconds for models with more than 15,000 equations. Consequently, an average-sized model will simulate a scenario from 1980 to 2050 in less than a second and provide decision makers with immediate feedback
- **Realistic insights**. A causal descriptive model can capture new patterns of behaviour and help identify potential side-effects and additional synergies

#### A5.2.2 Weaknesses

- **Demanding validation**. Although the statement 'garbage in–garbage out' does not apply to causal descriptive models (because data inputs are not the primary driver of the simulation's results), validation has to be carried out at two levels: the structural and the behavioural
- **Projections, not predictions**. As with any other methodology and model, simulations should be used for the trend they generate and a sense of the direction the system will take when interventions are implemented (e.g. which feedbacks change dominance and gain strength over time), rather than for the indicators' specific value at a specific point in time
- **A simplified representation of reality, not reality**. Simulation models are a simplified version of reality and, as such, need to be carefully built and tested. Defining the boundaries of the model

is therefore crucial to ensure that it represents the essential characteristics of reality and is not too large or too simple

## A5.3 Associated decision-making steps

Simulation can be utilised to effectively support strategy/policy assessment (identifying intervention options that will have the desired impact on the system and the right degree of impact), as well as evaluation (simulating selected intervention options).

Simulation models are most commonly used when the analysis is done *ex ante*, or before the actual implementation of the interventions.

- *Ex ante* modelling can generate 'what if' projections of the expected (and unexpected) impacts of proposed strategy/policy options on a variety of key indicators across sectors. In addition, well-designed models, which merge economic and biophysical variables and sectors, can help with the cost–benefit analysis and the prioritisation of strategy/policy options. The use of structural models that explicitly link interventions with their impacts can generate effective projections of how a certain target could be reached and when

On the other hand, an *ex post* analysis could also be useful to monitor trends and analyse why projections may differ from reality in the months and years after the strategy/policy implementation.

- *Ex post* modelling can support impact evaluation by improving the understanding of the relations between the key variables in the system and by comparing the projected performance with the initial conditions and historical data. Improvements to the model and updated projections allow decision makers to refine the targets and objectives, as well as to build on synergies and positive spillovers across sectors.

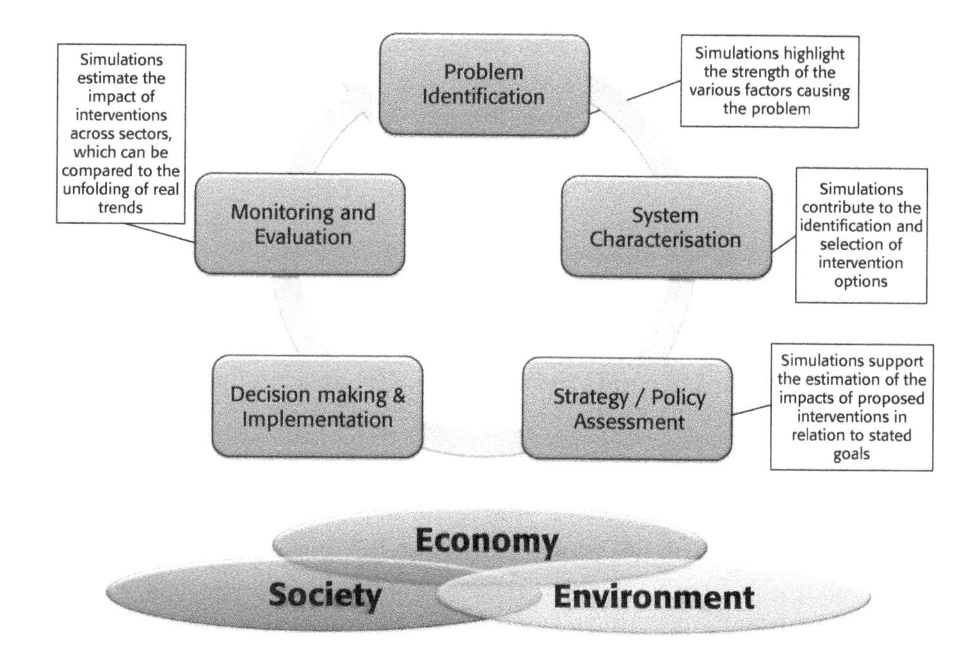


FIGURE 22 Simulation to inform the integrated decision-making process

## A5.4 Possible combination with other tools

A simulation carried out with system dynamics (SD) models—normally cross-sectoral simulations—extends the analysis done with indicators (which are included in the model as the main variables) and makes use of influence tables, as well as the CLD, to correctly represent the analysed problem characteristics (and system). Simulation is a natural extension of the creation of CLDs, which use systems thinking (ST). In practice, SD is a methodology to quantify ST. Consequently, by also using scenarios to define what to simulate and analyse, simulation integrates all the tools proposed in this book into a single analysis framework.

## A5.5 Implementation steps

The development of a system dynamic model, including conceptualisation, customisation and validation, proceeds through a variety of tasks, and the generic steps are presented below:

1. **Identification of key issues and opportunities**. As every model application is unique, the issues to be analysed have to be carefully demarcated and agreed upon. A multi-stakeholder process is often adopted to obtain the widest possible stakeholder views
2. **Data collection and consistency check**. This is a time-consuming task and, besides data mining, cross-sectoral data consistency checks are an essential step
3. **Causal mapping and identification of feedback loops**. This step constitutes creating causal loop diagrams (CLDs) as described previously
4. **Creation of customised mathematical models**. This step is a sequence of iterations involving the key stakeholders. It comprises translating CLDs into mathematical models with numerical inputs and equations. At this stage, the model is built on social, economic and environmental sectors. In practice, it integrates the best sectoral knowledge into a single model framework. This framework represents the full incorporation of the economic and biophysical variables that captures: a) feedbacks within and across sectors, which is aimed at identifying synergies and potential bottlenecks (unexpected side-effects); b) the time delays, whereby policies and investment allocations may lead to a 'worse before better' situation; and c) the nonlinearity, leading to the identification of potential thresholds and tipping points
5. **Validation and analysis**. Variables and equations have to be validated to ensure that all experts feel comfortable with the model's overall structure and that it reflects the reality observed. This is primarily done by simulating the base case and testing the outputs of the simulations against historical data on a multitude of socioeconomic and environmental indicators. Decision makers' confidence that the causal relations in the model are well established emerges from the model's ability to replicate historical

data. Where necessary, the model can be calibrated to obtain a consistent and reliable baseline simulation—the business-as-usual (BAU) case

6. **Simulation of alternative scenarios**. Once the BAU is confirmed, scenarios can be simulated to test the impacts of alternative strategy/policy options, which were identified in Step 1

# References


Agrawala, S., M. Carraro, N. Kingsmill, E. Lanzi, M. Mullan and G. Prudent-Richard (2011) *Private Sector Engagement in Adaptation to Climate Change* (Paris: OECD Publishing).

Annacchino, M. (2003) *New Product Development: from Initial Idea to Product Management* (USA: Elsevier).

Bassi, A. (2009) *Systems Modeling of Long Term Energy Policy, Mauritius* (Port Louis, Mauritius: Ministry of Renewable Energy and Public Utilities, Republic of Mauritius and UNDP Country Office Mauritius and Seychelles).

Bassi, A., and R. Mistry (2009) 'Assessing Water Management Options on Maui: Applying an Integrated Approach to Inform Community Conversations and Policy Debates', *Environmental Science and Engineering Magazine*, Fall Issue.

Bassi, A., and J. Yudken (2009) 'Potential Challenges Faced by the U.S. Chemicals Industry Under a Carbon Policy', *Sustainability* 1: 592-611.

Bassi, A., K. Balnac, C. Bokhoree and P. Deenapanray (2009a) 'A System Dynamics Model of the Mauritian Power Sector', *Proceedings of the 27th International Conference of the System Dynamics Society*, Albuquerque, NM, 26–30 July 2009.

Bassi, A., J. Harrison and R. Mistry (2009b) 'Using an Integrated Participatory Modeling Approach to Assess Water Management Options and Support Community Conversations on Maui', *Sustainability* 1.4: 1331-48.

Bassi, A., J. Yudken and M. Ruth (2009c) 'Climate Policy Impacts on the Competitiveness of Energy-Intensive Manufacturing Sectors', *Energy Policy* 37: 3052-60.

Bassi, A., H. Herren, Z. Tan and B. Saslow (2011) 'Assessing Future Prospects of the Agriculture Sector Using an Integrated Approach', *Agricultural Science Research Journal* 1.4: 92-101.

Brehmer, B. (1992) 'Dynamic Decision Making: Human Control of Complex Systems', *Acta Psychologica* 81.3: 211-41.

Brown, S., and H. Huntington (2008) 'Energy Security and Climate Change Protection: Complementarity or Tradeoff?' *Energy Policy* 36.9: 3510-13.

Büchel, B., and G. Probst (2001) *Deutsche Bank: Measuring Employee Retention* (Lausanne, Switzerland: IMD International).

Butland, B., S. Jebb, P. Kopelman, K. McPherson, S. Thomas, J. Mardell and V. Parry (2007) *Tackling Obesities: Future Choices—Project Report* (London: Foresight, UK Government Office for Science).

Centre for Environmental Systems Research (2009) *GLOWA Jordan River Scenarios of Regional Development under Global Change* (Kassel, Germany: Centre for Environmental Systems Research, University of Kassel).

Churchman, C. (1967) 'Guest Editorial: Wicked Problems', *Management Science* 14.4: 141-42.

CNA Corporation (2007) *National Security and the Threat of Climate Change* (Alexandria, VA: CAN Corporation).

Damer, T. (2009) *Attacking Faulty Reasoning: A Practical Guide to Fallacy-Free Arguments* (Stamford, CT: Cengage Learning).

Dorner, D. (1980) 'On the Difficulties People Have in Dealing With Complexity', *Simulation and Games* 11.1: 87-106.

EIA (Energy Information Administration) (2009) *International Energy Statistics* (Washington, DC: US EIA).

Einhorn, H., and R. Hogarth (1986) 'Judging Probable Cause', *Psychological Bullettin* 99: 3-19.

FAO (Food and Agriculture Organisation) (2008) *The State of World Fisheries and Aquaculture* (Rome: FAO).

FAO (2009a) *Report of the FAO Expert Meeting on How to Feed the World in 2050* (Rome: FAO).

FAO (2009b) *FAOSTAT* (Rome: FAO).

FAO (2009c) *The State of World's Forests* (Rome: FAO).

Fiksel, J. (2006) 'Sustainability and Resilience: Toward a Systems Approach', *Sustainability: Science, Practice & Policy* 2.2: 14-21.

Financial Executives International (FEI) (2001) *Ensuring Better Business Forecasting* (Executive Briefing, December 2001; Morristown, NJ: FEI).

Forrester, J.W. (1961) *Industrial Dynamics* (Portland, OR: Productivity Press): 14.

Gall, J. (1977) *Systemantics: How Systems Work and Especially How They Fail* (New York: Quadragle/The New York Times Book Co).

Global Forest Coalition (2008) *Life as Commerce: The Impact of Market-Based Conservation on Indigenous Peoples, Local Communities and Women* (Asunción, Paraguay: Global Forest Coalition).

Government of Qatar and UNDP (2009) *Qatar National Vision 2030. Advancing Sustainable Development. Qatar's Second Human Development Report* (Doha: Government of Qatar and UNDP).

Heinhorn, H., and R. Hogarth (1978) 'Confidence in Judgment: Persistence of the Illusion of Validity', *Psychological Review* 85.5: 395-476.

Hemmati, M., F. Dodds and J. Enayati (2002) *Multi-stakeholder Processes for Governance and Sustainability* (London: Earthscan).

Hofstede, G., and M. Bond (1988) 'The Confucius Connection: From Cultural Roots to Economic Growth', *Organizational Dynamics* 16.4: 5-21.

Howarth, R., and P. Monahan (1996) 'Economics, Ethics and Climate Policy: Framing the Debate', *Global and Planetary Change* 11.4: 187-99.

Huang, W., D. Cheng and J. Dagsvik (2012) 'The Impact of Price on Chemical Fertilizer Demand in China', *Asian Agricultural Research* 4.7: 7-12.

Hurst, P. (2006) *Agriculture Workers and Their Contribution to Sustainable Agriculture and Rural Development* (Geneva: ILO).

IAASTD (International Assessment of Agricultural Knowledge, Science and Technology for Development) (2009) *Agriculture at a Crossroad (Agricultural Assessment)* (Washington, DC: IAASTD).

IISD (International Institute for Sustainable Development) (2013) *A Guidebook to Fossil-Fuel Subsidy Reform for Policy-Makers in Southeast Asia* (Winnipeg, Manitoba: IISD).

IPCC (Intergovernmental Panel on Climate Change) (2007) 'Summary for Policymakers', in IPCC, M. Parry, O. Canziani, J. Palutikof, P. van der Linden and C. Hanson (eds.), *Climate Change 2007: Impacts, Adaptation and Vulnerability. Contribution of Working Group II to the Fourth Assessment Report of the Intergovernmental Panel on Climate Change* (Cambridge, UK: Cambridge University Press): 7-22.

Kennedy, E., V. Mannar and V. Iyengar (2003) 'Alleviating Hidden Hunger: Approaches That Work', *IAEA Bulletin* 45.1.

Maack, J. (2001) *Scenario Analysis: A Tool for Task Managers* (Washington, DC: World Bank Group).

McKinsey & Company and 2030 Water Resources Group (2009) *Charting Our Water Future* (Washington, DC: 2030 Water Resources Group).

Miller, J., and S. Page (2007) *Complex Adaptive Systems: An Introduction to Computational Models of Social Life* (Princeton, NJ: Princeton University Press).

Muller, A., and J. Davis (2009) *Reducing Global Warming: The Potential of Organic Agriculture* (Policy Brief, no.31.5.2009; Kutztown, PA: Rodale Institute).

Nitkin, D., R. Foster and J. Medalye (2009) *Case Studies & Tools. A Systematic Review of the Literature on Business Adaptation to Climate Change* (London, Ontario: Network for Business Sustainability).

Nutt, P. (2002) *Why Decisions Fail: Avoiding the Blunders and Traps that Lead to Debacles* (San Francisco: Berrett-Koehler Publishers).

Ogwueleka, T. (2009) 'Municipal Solid Waste Characteristics and Management in Nigeria', *Iranian Journal of Environmental Health Science & Engineering* 6.3: 173-80.

Probst, G. (1985) 'Some Cybernetic Principles for the Design, Control and Development of Social Systems', *Cybernetics and Systems, An International Journal*: 277.

Probst, G., and P. Gomez (1988) 'Pitfalls of Managerial Thinking in Complex Systems', in R. Trappl (ed.), *Cybernetics and System Research* (Dordrecht: Reidel): 1151.

Probst, G., and P. Gomez (1989) 'Thinking in Networks to Avoid Pitfalls of Managerial Thinking', *Human Systems Management* 8.3: 201.

Probst., G., and P. Gomez (1995) *Die Praxis des ganzheitlichen Problemlösens* (Bern: Haupt).

Probst, G., S. Raisch and F. Ferlic (2008) 'Unternehmerische Balance: Nestlés organisches Wachstum in reifen Märkten', *Zeitschrift für Führung und Organisation* 3: 170-76.

Richardson, G., and A. Pugh (1981) *Introduction to System Dynamics with Dynamo* (Portland, OR: Productivity Press).

Roberts, N., D. Andersen, R. Deal, M. Garet and W. Shaffer (1983) *Introduction to Computer Simulation. The System Dynamics Approach* (Reading, MA: Addison-Wesley).

Schneider, J., and J. Hall (2011) 'Why Most Product Launches Fail', *Harvard Business Review*, April 2011.

Serrie, H. (1986) 'Anthropology and International Business', in G. Ferraro (ed.), *The Cultural Dimension of International Business* (Upper Saddle River, NJ: Pearson Prentice Hall).

Seychelles Fishing Authority (SFA) (2012) *Seychelles Tuna Bulletin. Year 2011* (Mahe, Seychelles: SFA).

Shimizu, K. (2000) *Transforming Kaizen at Toyota* (Okayama, Japan: Okayama University).

Siemens AG (2013) 'Megatrends', Siemens, https://www.swe.siemens.com/belux/web/en/about/megatrends/Pages/megatrends.aspx, accessed 10 October 2013.

Simon, H. (1982) *Models of Bounded Rationality* (Cambridge, MA: MIT Press).

South African Food Pricing Monitoring Committee (FPCM) (2003) *Analysis of Selected Food Value Chains. Part 4 of the 2003 FPCM's Final Report* (Pretoria: FPCM).

Stadtler, L., and G. Probst (2013) *Planting the Seeds of Change: The Ethiopia Commodity Exchange* (Geneva, Switzerland: DHEC - University of Geneva).

Stadtler, L., and L. van Wassenhove (2011) *Building on Lessons Learnt: Disaster Relief Operations at Agility* (Fontainebleau, France: INSEAD).

Stadtler, L., T. Arabiyat and G. Probst (2010) *Creating Shared Responsibility in a Multi-Stakeholder Partnership* (Geneva, Switzerland: HEC - University of Geneva).

Sterman, J.D (2000) *Business Dynamics: Systems Thinking and Modeling for a Complex World* (Boston, MA: Irwin/McGraw-Hill).

TEEB (The Economics of Ecosystems and Biodiversity) (2010) *TEEB for Business: Executive Summary* (Geneva: TEEB).

Tversky, A., and D. Kahneman (1974) Judgment under Uncertainty: Heuristics and Biases', *Science* 185.4157: 1124-31.

UNDP (United Nations Development Programme) (2009) *Handbook on Planning, Monitoring and Evaluating for Results* (New York: UNDP).

UNDP (2011) *Updated Guidance on Evaluation in the Handbook on Planning, Monitoring and Evaluation for Development Results* (New York: UNDP).

UNDP (2012) *Multi-Stakeholder Decision-Making: A Guidebook for Establishing a Multi-Stakeholder Decision-Making Process to Support Green, Low-Emission and Climate-Resilient Development Strategies* (New York: UNDP Publication).

UNEP (United Nations Environment Programme) (2009) *Integrated Policymaking for Sustainable Development, A Reference Manual* (Geneva: UNEP).

UNEP (2010a) *Assessing the Environmental Impacts of Consumption and Production: Priority Products and Materials. A Report of the Working Group on the Environmental Impacts of Products and Materials to the International Panel for Sustainable Resource Management* (Paris: UNEP Publication).

UNEP (2010b) *Risk and Vulnerability Assessment Methodology Development Project (RiVAMP). Linking Ecosystems to Risk and Vulnerability Reduction. The Case of Jamaica. Results of the Pilot Assessment* (Geneva: UNEP).

UNEP (2011a) *Towards a Green Economy: Pathways to Sustainable Development and Poverty Eradication* (Nairobi: UNEP).

UNEP (2011b) *Green Economy Success Stories from the UNECE Region* (Geneva: UNEP).

UNEP (2012a) *Global Environment Outlook 5* (Geneva: UNEP).

UNEP (2012b) *Measuring Progress Towards an Inclusive Green Economy* (Geneva: UNEP).

UNEP (2013) *Green Economy and Trade: Trends, Challenges and Opportunities* (Geneva: UNEP).

UNEP, ILO, IOE and ITUC (2008) *Green Jobs: Towards Decent Work in a Sustainable, Low-carbon World* (Geneva: UNEP).

USAID (United States Agency for International Development) (2012) *Multi-Stakeholder Evaluation of Agriculture and Livestock Value Chain Activities in Kenya* (Washington, DC: USAID).

Van Paddenburg, A., A. Bassi, E. Buter, C. Cosslett and A. Dean (2012) *Heart of Borneo: Investing in Nature for a Green Economy* (Jakarta: WWF Heart of Borneo Global Initiative).

Vester, F. (1978) *Urban Systems in Crisis. Understanding and Planning Human Living Spaces: The biocybernetic approach* (Stuttgart: Deutsche Verlags Anstalt).

Vester, F., and A. von Hesler (1982) *Sensitivity Model* (Frankfurt/Main: Umlandverband Frankfurt).

World Bank (2008) *World Development Report 2008: Agriculture for Development* (Washington, DC: World Bank).
World Bank (2013) 'Data', World Development Indicators, http://data.worldbank.org/data-catalog/world-development-indicators, accessed 5 April 2013.
World Economic Forum (2009) *Mining & Metals. Transparency and Ethics Dialogue* (Cape Town, South Africa: World Economic Forum).
World Economic Forum (2010) *Realizing a New Vision for Agriculture: A Roadmap for Stakeholders* (Geneva: World Economic Forum).
World Economic Forum (2012) *Global Risks 2012. Seventh Edition* (Geneva: World Economic Forum).
World Economic Forum (2013) *Multistakeholder Collaboration for Healthy Living: Toolkit for Joint Action* (Geneva: World Economic Forum).
Worm, B., E. Barbier, N. Beaumont, J. Duffy, C. Folke, B. Halpern, J.B.C. Jackson, H.K. Lotze, F. Micheli, S.R. Palumbi, E. Sala, K.A. Selkoe, J.J. Stachowicz and R. Watson (2006) 'Impacts of Biodiversity Loss on Ocean Ecosystem Services', *Science* 314: 787-90.
Yudken, J., and A. Bassi (2009a) 'Climate Change and US Competitiveness', *Issues in Science and Technology*, Fall Issue.
Yudken, J., and A. Bassi (2009b) *Climate Policy and Energy-Intensive Manufacturing: The Competitiveness Impacts of the American Energy and Security Act of 2009* (Arlington, VA: High Road Strategies; Washington, DC: Millennium Institute).
Zimmermann, A., P. Gomez, G. Probst and S. Raisch (2013) *Synergistic Value: How Smart Firms Create Public Value by Growing Their Core Business* (unpublished manuscript; Switzerland: Center for Leadership and Values in Society, University of St Gallen and University of Geneva).

# About the Authors

**Gilbert Probst** is Managing Director, Leadership Office and Academic Affairs, and Dean of the Global Leadership Fellows programme at the World Economic Forum. He is also a full professor for Organisational Behaviour and Management and co-director of the Executive MBA programme at HEC, University of Geneva, Switzerland. He has a PhD as well as habilitation in business administration from the University of St Gallen, Switzerland. As a visiting faculty member, Professor Probst taught at the Wharton School of the University of Pennsylvania, Philadelphia, as well as at the International Management Institute (IMI, merged into IMD) in Geneva. He also served as the president of the board of Swiss Top Executive Training (SKU), the Swiss Board Institute and as a member of the board of the Swiss Management Society. He is the founder of the Geneva Knowledge Forum as well as CORE (Center for Organizational Excellence) at the Universities of St Gallen and Geneva, and the Strategic Knowledge Group, Zürich. He is an award-winning author of several books, papers and cases and has served as a board member and consultant for several companies.

Dr **Andrea M. Bassi** is the founder and CEO of KnowlEdge Srl, a consulting company exploring socio-economic and environmental complexity to inform decision making for sustainability. He is also an Extraordinary Professor at Stellenbosch University in South Africa. In his work Dr Bassi is a project leader and researcher with over 10 years of experience supporting more than 20 governments, several international organisations and business leaders primarily on (1) green economy and green growth strategies, (2) action plans for resilience and risk mitigation, and (3) sustainable development planning. Dr Bassi's strengths lie primarily in the use of modelling and research techniques that focus on systemic analysis. He was a Director at Millennium Institute, and has worked across disciplines with the Danish National Environmental Research Institute and the Los Alamos National Laboratory. Dr Bassi holds a PhD and M.Phil in System Dynamics from the University of Bergen, Norway, and a MSc in Business and Economics from LIUC, Italy.

# Index